THE CARBON CREED

THE CARBON CREED

7 PATHWAYS TO ZERO

WALTER L. McLEOD

Paperback ISBN: 979-8-9868448-0-0
eBook ISBN: 979-8-9868448-1-7

Library of Congress Control Number: 2022921569

Cover Image: The abstract image features seven connected circular shapes that symbolize each step in the creed pathway. These organic forms, inspired by waterdrops, intersect and flow together to demonstrate the connected nature of our beliefs and actions with the environment. It invites you to consider your carbon truth.

*To the two who gave me life
and the ones who make a difference*

Contents

INTRODUCTION

*"If a man writes a book, let him set down only what he knows.
I have guesses enough of my own."*
—JOHANN WOLFGANG VON GOETHE

In January 2020, something eerie and unnerving happened in the Republic of Sakha in Eastern Siberia. During the dead of winter with a recorded temperature of minus fifty degrees Celsius (minus fifty-eight degrees Fahrenheit), white smoke started to rise from the frozen ground. Observers were stunned to learn that the source of the smoke was a peat fire beneath the snow-covered ground. Scientists call them "zombie fires," and they are thought to be caused by global warming. Blanketed by snow, these fires hibernate underground, smoldering through the winter. When the ground thaws as spring arrives, the fires reemerge on the surface fueled by carbon-rich peat and Northwoods soils. Scientists expect zombie fires to burn for years to come.

Just four years earlier, people of Western Siberia experienced an equally dystopian crisis—a phantom anthrax outbreak. The outbreak occurred on the Yamal Peninsula, high above the Arctic Circle at the top of the world. The source of the outbreak was the melted corpse of a reindeer that had been frozen for more than seventy-five years. One boy and over 2,000 reindeer died as a result of the outbreak. Scientists believe that global warming caused the corpse to thaw.

For anyone familiar with the Artic, this all seems impossible. The Yamal region is so cold the soil, called permafrost, is frozen solid more than 1,000 feet deep in many places, about the height of the Eiffel Tower in Paris. Extinct animals and people have been buried in permafrost for centuries. There could be human bodies infected with all kinds of viruses and bacteria frozen in time. In recent years, researchers have unearthed traces of the 1918 Spanish flu virus in corpses buried nearly a century ago in Alaska's tundra. It is possible that smallpox and bubonic plague are also buried in Siberia. All of this has increased concern among scientists about the likelihood of more dormant pathogens reactivating and potentially spreading as the earth continues to warm. As we've witnessed with the COVID-19 pandemic, a highly contagious virus has no national borders and can quickly derail human society.

When we read these accounts and soak them in, they can be very unsettling. But we need to be uncomfortable with this. The earth is sick, and we need to find a cure. Siberia is just our canary in a coal mine—an early warning of the havoc climate change is bringing to human life on every continent. It doesn't have to be that way. Human beings are resilient, and we have the know-how and technology that can curb a changing climate. The question is, will we? I believe that humanity can successfully navigate the climate crisis if we act with urgency and purpose.

How I Came To Write This Book

When I started my career in the early 1990s, most people viewed the environment as a fringe issue. The phrase "tree hugger" was hurled as a slur by some. Energy was synonymous with wealth and power, so I felt quite lucky landing a job in the energy industry in my mid-20s. For the next eight years, I sat at the table with some of Washington's leading minds on energy and environment policy. It was clear to me that working for the energy industry had given me an opportunity and a platform to shape the nation's energy policy that was rare for a person of my age and background—a young, African American man. It was audacious!

However, it was during that time that I started asking myself hard questions about the long-term viability of the petroleum industry and its impacts on the climate. How and when would we begin transitioning to a lower-carbon economy? What role did I want to play in that transition? To further complicate matters, my wife had just given birth to our daughter, causing me to think about the world she might inherit in a few decades and my part in shaping it.

Although I didn't realize it at the time, I was starting down my personal road to Damascus. Like the biblical Saul of Tarsus, I was about to undergo a conversion that would change the trajectory of my life and career in ways I could never have imagined.

In 1999, to the surprise of my colleagues and mentors, I left the energy industry to start a nonprofit organization called Clean Beaches Council. I was a clueless thirty-three-year-old trying to run a start-up with no business partners or experience. That is the recipe for failure, my friends. Except that's not what happened.

Within a few years my robust board of directors, advisers, and I raised more than one million dollars and put together one of the most effective coastal policy campaigns of our time. In 2007, supported by a small staff of dedicated interns, I created a coastal creed that was endorsed by the White House and over 150 governors, mayors, and counties. The creed became a House and Senate resolution that was passed by unanimous consent in the 110th US Congress. Today it's known as National Clean Beaches Week and is observed annually during the first seven days of July.

Like Saul who became Paul, this was my conversion moment. I was different. I had changed. Finally, I had found a way to align my personal beliefs, values, and practices with my work life.

Since then I have been living my truth when it comes to carbon and climate. I run businesses that develop and finance solar farms, invest in clean technology, and shape public policy. Now I can add author to that list. I'm not a journalist, so this book isn't likely to win praiseworthy reviews from critics, but I didn't write it for the critics. I'm not a practicing scientist, but I was trained as a chemist some decades ago. I think that counts for something. I am, however, someone who has worked professionally in the world of energy and the environment for the past thirty years—and a lot has changed, including me.

Why am I sharing my story with you? Because you, too, can live your carbon truth. There's no reason you cannot have the sustainable, healthy, and prosperous life you've always wanted, and you don't have to quit your job or write legislation to do it. I wrote this book to show you how.

What's In This Book

This book is organized around seven affirmations of the Carbon Creed. The creed is designed to help you identify and adopt the beliefs, values, and practices that lead to decarbonization. Below is a brief synopsis of each chapter:

Chapter 1: Beliefs.

What do you believe? I will introduce the core beliefs of the Carbon Creed. Using positive affirmations, you will learn how to train your subconscious mind to achieve your carbon goals.

Chapter 2: Ethics.

We'll take a deep dive into the concept of ethical rules. Human society has always been shaped by ethical norms, and climate change is no different. We will draw lessons from the teachings and experiences of Confucius, Abraham, and two tech titans.

Chapter 3: Mindset.

We will dissect the lives of two historic figures who shared similar ecological values but followed very different paths. You will discern your own mindset and leanings toward human innovation and conservation.

Chapter 4: Science.

We will discuss the history of our carbon addiction and how humans have become the dominant force on Earth. We round it out with a hopeful discussion on overcoming climate anxiety and emerging technologies that could mitigate carbon emissions.

Chapter 5: Kinship.

We will discuss the importance of effective communication when engaged in the climate debate. Facts, while important, are not the key to changing people's hearts and minds. We'll reveal the secret to solving this problem, and it's not what you think.

Chapter 6: Habits.

You will learn why carbon and climate rituals are important and how they can help you put the creed into practice. I will reveal the five carbon-conscious rituals that unlock the door to decarbonize your life. The great poet Maya Angelou will provide inspiration.

Chapter 7: Accountability.

I will explain the narrative behind carbon pledges and provide the framework for individuals like you to hold governments and corporations accountable. You are more powerful than you think.

Chapter 8: Equity.

I will share my view that equity, in the context of climate, has two sides—social and economic. You will learn why we need both sides to achieve a fair, sustainable future for everyone. Martin Luther King Jr. and the fictional characters Lieutenant Uhura and Yoda play an instructive role in our discussion.

How To Experience This Book

This is not a book about the science of climate change, nor is it a policy brief. It's not meant to scare you, though some data might.

This book is written to take you on a journey through the climate narratives nestled in your head and heart. Along the way, you will begin to answer the fundamental question: What do I believe?

You may think that's a simple question, but I assure you it's not. What would *other people say you believe* about carbon and climate? What vibe are you sending in word and deed? Do you virtue signal, engaging in moral grandstanding to impress others? Or do you live your truth whether you get accolades or attention or not? Does it depend on who's in the room? Living your carbon truth is more than a label. It's more than donating to a worthy cause or being named at the top whatever list. What you believe about climate is your Carbon Creed.

As you read and meditate on the principles and practices of the Carbon Creed, your climate narrative will emerge. Metaphors are a potent tool to describe the carbon and climate crisis. The two most common metaphors used in the media are war and health. Usually when people talk about the war on climate, they are calling on the masses to mobilize against rising carbon emissions. People become angry. They take sides. Climate change, however, is not evil, and there are no armed forces waiting to defeat it, so I do not recommend war as your climate narrative. Health is a more effective metaphor. We all want and need good health. It is the most valuable thing we can possess. We all need clean air, clean water, and robust forests to maintain our good health. But the earth is not well, and the symptoms of her illness are evident—just observe the weather. With every passing year and every changing season, we experience more severe weather events. Climate change knows no national boundaries, much like a global health pandemic.

Still, we must not lose hope. That's the purpose of this book—to give you hope. It will help you confront, consider, and adopt beliefs, values, and practices that will inspire you to live your carbon truth. You are the cure.

CHAPTER 1
BELIEFS

Today is the beginning of your carbon and climate journey. It starts with the belief that we are living in the Anthropocene, a time when humans are the dominant force on Earth, terraforming and changing the planet's ecosystems and processes. This has led to human-induced climate change—the unsustainable warming of the planet caused by burning hydrocarbons that release greenhouse gases. Some people refuse to acknowledge the threat and urgency of the carbon problem—to them climate change feels overwhelming and defeating. But we refuse to accept that narrative. We believe that each of us has a role in changing the narrative from fear to courage—that we all have a shared obligation to live a carbon-conscious life.

The Carbon Creed is your foundation on this journey. It is a pathway to the beliefs, values, and practices that will inspire you and others to decarbonize. Many people are on this journey, each with a different story and motive. They are a mixed band of pilgrims, each playing a part in the *Canterbury Tales* of climate change. They include:

- *the employee* who wants to see his company do more to reduce its carbon footprint;

- *the consumer* who feels strongly about decarbonizing her purchases or daily activities;

- *the investor* who wants his portfolio companies to adopt transparent environmental, social, and governance (ESG) reporting standards;

- *the advocate* who wants the government to address climate-induced health disparities in her community;

- *the spiritual person* who wants his faith to better reflect his carbon consciousness;

- *the parent* who wants to instill a sense
  of carbon stewardship and awareness
  in her child.

Do you see yourself in any of these characters? Do you share their stories?

Ultimately, this narrow list of climate pilgrims is not what matters. What matters is that you chose this path because you understand and accept the need to act on carbon and climate. You chose this path because you are ready to adopt the creed.

Why A Creed?

The word *creed* comes from the Latin word *credo*, meaning "I believe." It is the first word of many religious creeds, such as the Apostles' Creed. But it can also be applied to any guiding principle or set of principles, like the American Creed first formulated by Thomas Jefferson. Simply put, a creed is a set of beliefs or aims that guide someone's actions. Creeds are important for a number of reasons:

- Creeds summarize our beliefs. They provide a consensus of the basic beliefs a person or group holds at any given time.

- Creeds are timeless and help us respond to the challenges of the day. They restate essential beliefs that have been challenged and reveal how we should respond.

- Creeds affirm what we believe and why we believe it. They can reinforce positive beliefs and messages into our subconscious mind.

- Creeds unite all believers across time. They serve as a reminder that we are connected across past, present, and future generations.

- Creeds provide an ethical foundation for life. Because they express our values and beliefs, creeds can be a helpful reference for making daily decisions.

The Carbon Creed must embody all these qualities, unifying humanity to successfully address our underlying hydrocarbon addiction and the crisis it causes to our climate. You need to train your mind to become more conscious of carbon and climate. Specifically, your subconscious mind.

The mind is the seat of thought. It functions in two ways—conscious and subconscious. According to scientific research, your conscious mind makes up less than 10 percent of your total brain function. That means that the subconscious aspect of your mind is around 90 percent of your total brain function. What most people don't realize is that you can train your subconscious mind to help you achieve exactly what you desire in life. The most effective way to accomplish this is through repetitive, positive affirmations. With time and diligence, affirmations can become self-fulfilling prophecies.

According to neuropsychiatrist Dr. Mona L. Schulz, repeating affirmations physically changes the neural pathways in your brain. Affirmations are a highly effective way to train your subconscious mind and act on a new belief you have formed. For example, an affirmation can be as simple as, "I believe in myself" or "I will not worry about things I cannot control."

The Carbon Creed is written as a series of positive affirmations. When you say "I believe," you are making a positive affirmation. If you are serious about achieving your carbon and climate goals, then study and absorb the Carbon Creed to train your subconscious mind to produce the outcomes you want in your life.

The Carbon Creed

i. I believe all carbon virtues and values can be expressed in two ancient rules:

> <u>Do no harm</u>. In accordance with the Confucian tradition, we should put the welfare of "others" before our "self." This is the Silver Rule.

> <u>Do good</u>. In accordance with the Abrahamic tradition, we should do good to others when we have the ability. This is the Golden Rule.

ii. I believe a sustainable human society requires both innovation and conservation.

iii. I believe in the scientific consensus that the earth's climate is warming and human activities are the primary cause.

iv. When seeking common ground about carbon and climate change, I believe it is more effective to ask and listen before speaking.

v.   I believe transforming everyday habits and routines into carbon-conscious rituals gives meaning and purpose to life.

vi.   I believe governments and companies are ultimately responsible for the impact of greenhouse gas emissions on human health and society.

vii.   I believe in climate equity and opportunity for all humans.

The Affirmations

Affirmations are encouraging words you tell yourself to help you reach your goals, overcome your fears, and achieve your purpose. Let's examine each of the seven creeds and find ways to apply them:

i. I believe all carbon virtues and values can be expressed in two ancient rules:

> <u>Do no harm</u>. In accordance with the Confucian tradition, we should put the welfare of "others" before our "self." This is the Silver Rule.

> <u>Do good</u>. In accordance with the Abrahamic tradition, we should do good to others when we have the ability. This is the Golden Rule.

This is the ethics creed. The Silver and Golden Rules are complimentary. Together they embody the moral imperative of the entire creed. If you start by doing no harm to others and do good when you have the ability, then your carbon virtues, values, and actions will be aligned. You will find the path to decarbonization.

---

ii.   I believe a sustainable human society
       requires both innovation and conservation.

This is the mindset creed. Innovation and conservation are complimentary mindsets for climate believers. Innovation expresses optimism in benevolent use of data and technology to arrest climate change and decarbonize human civilization. Conservation promotes carbon minimalism and embraces a shared, multigenerational responsibility to steward the earth.

---

iii.   I believe in the scientific consensus
        that the earth's climate is warming, and
        human activities are the primary cause.

This is the science creed. We live in the Anthropocene, a geological age when humans are the primary cause of planetary change. Use this affirmation to strengthen your belief that rising carbon emissions present an existential threat to human health, the economy, and the environment.

---

iv.   When seeking common ground about carbon and climate change, I believe it is more effective to ask and listen before speaking.

This is the kinship creed. Use this principle to discuss climate with people who do not necessarily share your convictions on carbon. Listen. Be kind. Find common ground. All humans are kin.

---

v.   I believe transforming everyday habits and routines into carbon-conscious rituals gives meaning and purpose to life.

This is the habits creed. Use this lens to elevate everyday habits and routines into meaningful rituals that decarbonize your life and engage your community.

---

vi. I believe governments and companies are ultimately responsible for the impact of greenhouse gas emissions on human health and society.

This is the accountability creed. Use this lens to assess governments' and corporations' commitments to adopt zero-carbon pledges, understand the metrics and data behind the pledges, and keep both groups transparent and accountable.

---

vii. I believe in climate equity and opportunity for all humans.

This is the equity creed. It has both social and economic dimensions that transcend race, wealth, nationality, gender, and political barriers. This affirmation allows you to support fair and equitable access to climate investments and economic opportunities for all demographic groups and ensure that the people least responsible for carbon emissions are not the most impacted by climate change.

Reflection: Beliefs

This chapter has introduced the idea that the path to a clean conscious on climate is applying the Carbon Creed in your life, both personally and professionally. How can you do that?

First, you can commit to *reading* the Carbon Creed every day. This will train your subconscious mind to be carbon conscious. Reading these positive affirmations will set the tone for your whole day—and if you do each day right you will do life right.

Second, you can *share* the Carbon Creed with others. The goal in sharing positive experiences and practices of the creed is to create "network effects"— the additional value an idea gains as more people use it.

Third, you can put the Carbon Creed into *practice.* It is possible to transform existing habits and routines into carbon-conscious rituals that give new meaning to your life. You can also adopt new practices that reflect your carbon values and help you become a more carbon-conscious individual.

Benchmark

You are at the beginning of your carbon and climate journey. The Carbon Creed is your foundation for that journey. The seven carbon creeds are:

- Ethics
- Mindset
- Science
- Kinship
- Habits
- Accountability
- Equity

Which of the carbon creeds do you feel most drawn to? Rank the creeds in order of importance to you today. Resolve to repeat the Carbon Creed affirmations every day, training your subconscious mind to produce the positive outcomes you want in life.

CHAPTER 2

ETHICS

"What you do not wish for yourself, do not do to others."
子貢問曰：”有一言而可以終身行之者乎”？子曰：”
其恕乎！己所不欲、勿施於人。
—CONFUCIUS

*"Society is defined not only by what it creates, but by what it
refuses to destroy."*
—JOHN SAWHILL

Most people are familiar with the Golden Rule of ethics. In the New Testament, Jesus summarized the spirit of the Mosaic law in a single phrase, "Do to others as you would have them do to you." However, the Golden Rule is not the first nor the only ethical rule, and there may be another better suited rule when it comes to addressing carbon and climate.

A Brief History of Ethical Rules

For all of human history, people have created rules and principles to define their relationship to each other. Dr. Yinya Liu, a professor and lecturer in Chinese philosophy, identifies the five historic rules of ethics that societies have used to define the relationship between the self and others.

First, is the ***Stone Rule***, which uses the principle "might makes right." This rule says the powerful can do whatever they please.

Second is the ***Iron Rule***, which promotes nationalism. This rule says conflict and competition between tribes is good (allies versus enemies).

Third is the ***Brass Rule***, which emphasizes the principle "an eye for an eye." This rule says take revenge (harm to those who harm you).

Fourth is the ***Silver Rule***, which is the principle of reciprocity. This rule says, "Do no harm."

Fifth is the ***Golden Rule***, which is the principle of benevolence. This rule says, "Do good."

Today, most humans seek to advance beyond the first three rules and consider the Silver and Golden Rules to be superior ethics.

FOLLOW THE SILVER RULE

Shu (Do no harm)

There is an important role for both the Silver and Golden Rules in modern society; however, when it comes to addressing climate change, which ethical rule should come first? To answer that, it helps to understand the cultural context that gave birth to the Silver Rule.

The most famous account comes from Confucianism, as recorded in the Analects. The Analects are a collection of the teachings and thoughts of Confucius dating to around 551–479 BC. They contain fragments of dialogues between the great Chinese philosopher and his disciples. In one such discussion, the disciple Zigong poses a question that gives rise to the Silver Rule.

Zigong (a disciple of Confucius) asks his Master: "Is there any one word that could guide a person throughout life?"

The Master (Confucius) replied: "How about 'shu:' never impose on others what you would not choose for yourself?"

—Confucius, Analects XV.24, tr.

In most versions, shu is translated as reciprocity, but there are many meanings behind this character. It also means empathy. Following the Confucian tradition you would prioritize others rather than yourself; you would meet the other person's needs first. The principle of putting "the other person's needs first" is paramount when applied to climate change. Many find it especially hard to make decisions and sacrifices that put the interests of future generations over their own.

Being respectful of the interests of past and future generations is key to the Confucian view of the self and groups. To the question, "Who am I?" the Confucian answers, "I am the child of my parents and the parent of my children."

Confucianism begins from the proposition that human beings are defined by kinship networks that span the centuries. From this perspective the interests of the individual are bound with the interests of the kinship group as it extends forward and backward across the generations.

Western cultures are beginning to see the value of this philosophy. Here is where the mindset creed and the ethics creed overlap. If we want a sustainable future, we need to embrace our shared, multigenerational responsibility to steward the planet. Apply the Silver Rule. Do no harm. Be a gracious descendant and ancestor.

TAKE THE HIPPOCRATIC OATH

(First, do no harm)

While the principle of *shu* may have origins in Asia, the Silver Rule is also embodied in a revered Western creed—the Hippocratic Oath. The Hippocratic Oath is an oath of ethics historically taken by physicians. The oath is the earliest expression of medical ethics in the Western world, establishing several principles of medicine that remain of paramount significance today. Most people recognize the Hippocratic oath from its seventeen-century summation, "First, do no harm" however, the oath dates back to AD 245. Below is the oath in its entirety:

I swear by Apollo Physician, by Asclepius, by Hygieia, by Panacea, and by all the gods and goddesses, making them my witnesses, that I will carry out, according to my ability and judgment, this oath and this indenture.

To hold my teacher in this art equal to my own parents; to make him partner in my livelihood; when he is in need of money to share mine with him; to consider his family as my own brothers, and to teach them this art, if they want to learn it, without fee or indenture; to impart precept, oral instruction, and all other instruction to my own sons, the sons of my teacher, and to indentured pupils who have taken the physician's oath, but to nobody else.

I will use treatment to help the sick according to my ability and judgment, but never with a view to injury and wrongdoing. Neither will I administer a poison to anybody when asked to do so, nor will I suggest such a course. Similarly, I will not give to a woman a pessary to cause abortion. But I will keep pure and holy both my life and my art. I will not use the knife, not even, verily, on sufferers from stone, but I will give place to such as are craftsmen therein.

Into whatsoever houses I enter, I will enter to help the sick, and I will abstain from all intentional wrongdoing and harm, especially from abusing the bodies of man or woman, bond or free. And whatsoever I shall see or hear in the course of my profession, as well as outside my profession in my intercourse with men, if it be what should not be published abroad, I will never divulge, holding such things to be holy secrets.

Now if I carry out this oath, and break it not, may I gain for ever reputation among all men for my life and for my art; but if I break it and forswear myself, may the opposite befall me.

—Translation by W.H.S. Jones (AD 245)

Although the exact phrase "First, do no harm" (Latin: *Primum non nocere*) does not appear in the original version of the oath, the intent is clear when it says, "I will abstain from all intentional wrongdoing and harm." The coronavirus pandemic (COVID-19) has revealed an important correlation between the Hippocratic Oath and the Silver Rule. We all now have a much greater appreciation for the sacrifice and service that health care providers make on behalf of society. Doctors, nurses, and EMTs all deserve our applause and gratitude for putting themselves on the front lines to save thousands of lives every day.

Let's also consider the impact of social distancing directives. When millions of people willingly sacrificed their freedom of movement (and income) by social distancing, millions of lives were saved. Collectively, we flattened the infection curve and tamed the health crisis. The same is true of wearing masks. We wear a mask primarily to protect others, not ourselves—to prevent the spread of the virus. How does this apply to the Silver Rule? We are not all doctors or nurses, but everyone can save lives by "doing no harm."

The same principle applies to climate change. We need masses of people to take the oath to "do no harm" if we ever want to flatten the carbon curve and tame the climate crisis.

THE FORCE

Another way to interpret shu is through the lens of science fiction. I am a huge science fiction fan. From Isaac Asimov's *Foundation* series to Frank Herbert's *Dune*, the genre invites us to consider the complex ways our choices and interactions contribute to generating the future.

Arguably the greatest science fiction story ever written is the *Star Wars* saga. George Lucas crafted an entire universe to shape our values through his epic work. In *Star Wars*, everything centers around the Force. According to Wikipedia, Lucas created the concept of the Force to "awaken a certain kind of spirituality" in young audiences. He developed the Force as a nondenominational religious concept, "distilled from the essence of all world religions," premised on the existence of God and distinct ideas of good and evil. That's why the Jedi meet in a temple and wear Buddhist-inspired robes.

There is an intersection between the idea of the Force and our discussion about the ethics creed. The Force is sometimes referred to in terms of light and dark sides. Lucas believed a character's intentions when using the Force—their "will to be selfless or selfish"—is what distinguishes light and dark sides. Thus, within the Force we find shu—do no harm. Recall that in the Confucian tradition, you would prioritize the other rather than the self; you would look at the other person's needs first. This is the Silver Rule. Rather than finding it difficult to place the interests of future generations over their own, the Jedi would understand this logic—and embrace it. Why is that?

The Jedi follow a special code of ethics: Jedi do not seize personal power or wealth; they seek knowledge and enlightenment. Jedi do not react to hatred, anger, fear, or aggression; they act when calm and at peace with the Force. Jedi use their powers to defend and protect, putting the needs of others first. In this way, the Jedi follow *shu*. As should you.

FOLLOW THE GOLDEN RULE

(Do good)

Over 2,000 years ago, Jesus of Nazareth summarized the whole of the Old Testament in a single phrase, "Do to others what you want them to do to you." This is the Golden Rule.

The Golden Rule has long been considered the cornerstone of the ethics of benevolence. It can be found almost in every cultural, ethical, and religious tradition, although differently formulated. According to scholar Daniel Esparza, the broad concept of "doing unto others" appears in some Middle Kingdom Egyptian papyri (2040–1650 BC), in the Code of

Hammurabi (1750 BC), in the writings of Thales (546 BC) and other Greek early philosophers, and in some Zoroastrian texts (300 BC–AD 1000). The Abrahamic religions (Judaism, Christianity, and Islam) also share the Golden Rule maxim. It is a concept that essentially no religion misses entirely—143 leaders of the world's major faiths endorsed the Golden Rule as part of the 1993 Declaration Toward a Global Ethic. However, belief in God or religion is not explicitly required to practice it.

In our discussion on carbon, we view the Golden Rule and Silver Rule as complimentary. Some believe they are essentially inverse expressions of the same thought. However, I believe that in the context of carbon and climate, there is greater urgency to first "do no harm" and then to "do good." It is more important (and potentially faster) to stop doing harm than to make amends.

Let's consider Elon Musk and his electric car company, Tesla, as an example. He is a rare and notable talent. His vision, hard work, wealth, and tenacity have made Tesla the model for decarbonizing transportation and the grid. We need people like Elon Musk to "do good." While he clearly has flaws, I believe Musk follows the Golden Rule. And yet, billions of people are still driving internal combustion engine (ICE) vehicles. Humans could decarbonize transportation quickly if everyone could afford electric vehicles. In the interim, if the masses committed to "do no harm" by taking a sabbath from ICE vehicles one day per week, the reduction in carbon emissions would be enormous. In that way the masses could follow the Silver Rule. Both rules are needed to reach the goal of a net-zero carbon society.

The Golden Rule reminds us, when it comes to climate, we should do good when we have the ability. Today there are many individuals who care about the climate and possess the resources and desire to do good on a massive scale. Jeff Bezos is one such person. Bezos began his transition from corporate titan to climate benefactor while leading Amazon, one of the most influential economic and cultural forces in the world. Consider a few of his climate accomplishments:

- In 2019, Bezos founded The Climate Pledge, an Amazon-led commitment to be net-zero carbon across its business by 2040, ten years ahead of the goal set by the Paris Agreement.

- Bezos led Amazon to become the biggest corporate buyer of renewable energy in the world, with a total of 127 solar and wind projects globally.

- Bezos created the Earth Fund in 2020, committing ten billion dollars of his personal wealth to funding research organizations, nonprofits, and others fighting the climate crisis.

In the summer of 2021, Bezos stepped down as CEO of Amazon, the company he founded in a rented garage in 1994. In making his announcement, Mr. Bezos listed three priorities he planned to focus on: (1) remaking *The Washington Post*, which he purchased in 2013; (2) exploring space with his rocket company Blue Origin; and (3) supporting organizations and individuals devoted to fighting climate change on Earth. Of these, fighting climate change will likely be his greatest contribution to humanity.

By establishing the Earth Fund, Jeff Bezos may be the greatest climate philanthropist in history. Despite his flaws, I believe Bezos follows the Golden Rule. He has effectively redefined the calculus and scale of what doing good for climate can be. To provide perspective, the Hewlett Foundation, which was the single largest private climate funder before Bezos, gives around $100 million per year. In 2020 the Earth Fund awarded its first sixteen grants totaling $791 million, five of which were $100 million each. That's nearly a sevenfold increase in annual climate funding.

Still, we should not become intoxicated with irrational exuberance for Bezos or any other tech titan. While the Earth Fund may be the greatest individual act of "good" for climate funding, it alone will not solve the problem. Ten billion dollars is a relatively small amount compared to the global hydrocarbon economy. Consider that the US government alone spends over two billion dollars per year on climate-related research funding. For humanity to transition the global economy away from fossil fuels, we need orders of magnitude more investment. This is where the masses come in. We need people like you to do good too. Ultimately, our choices as individuals are what matters most. People like Jeff Bezos can provide the spark, but we need the collective action and resources of individuals, corporations, and governments to leverage acts of benevolence into sustainable solutions. You don't have to be rich to do good.

Reflections: Ethics

This chapter has shown the wisdom found in ancient principles. Learn to use and apply the ethics creed as a compass to decarbonize your life. Understand your responsibilities as a descendant and ancestor in the stewardship of a sustainable planet. Embrace the Hippocratic Oath with a climate lens. Understand the lessons of the Force—be selfless and use any powers you possess for good. Always remember, the Silver and Golden Rules are complimentary. First, do no harm, then do good when you have the ability.

ETHICS CREED

I believe our carbon virtues and values can be expressed in two ancient rules:

<u>Do no harm</u>. In accordance with the Confucian tradition, we should put the welfare of "others" before our "self." This is the Silver Rule.

<u>Do good</u>. In accordance with the Abrahamic tradition, we should do good to others when we have the ability. This is the Golden Rule.

Benchmark

Humans have long created rules and principles to define their relationships to each other. The ethical rules necessary for humans to tackle carbon and climate change are the Silver and Golden Rules.

Which of these ethical rules speaks to you? Identify and document at least one carbon-conscious action you can take in the next four weeks to apply the ethics creed.

CHAPTER 3

MINDSET

"We are like tenant farmers chopping down the fence around our house for fuel when we should be using nature's inexhaustible sources of energy—sun, wind, and tide."
—THOMAS EDISON

"Electric power is everywhere present in unlimited quantities and can drive the world's machinery without the need of coal, oil, gas, or any other of the common fuels."
—NIKOLA TESLA

In a tweet on December 6, 2019, Elon Musk proposed turning 10,000 square miles in the US desert into a solar farm that could power the entire nation. "All you need is a 100 by 100-mile patch in a deserted corner of Arizona, Texas or Utah (or anywhere) to more than power the entire USA."

To check the math, University College London energy researcher Andrew Smith did some back-of-the-envelope calculations, and he found that a 10,000 square kilometer area—which is far smaller than the "100 by 100-mile patch"—could generate about 500 gigawatts, which is the average consumption in northwest Texas, at 21 percent efficiency. That's more than the 2013 US annual average consumption of 425 gigawatts, as reported by the US Energy Information Administration. People like Elon Musk are so talented in technology innovation that they seem to be able to achieve whatever they put their minds to.

While others, like marine biologist Dr. Sylvia Earle thrive in a similar cerebral zone that connects them to the natural environment in a Zen-like way. Referred to as "Her Deepness," Earle, a National Geographic Society Explorer-in-Residence, holds the record for deepest walk on the sea floor and is a world-renowned expert on marine biology. The first woman to lead the National Oceanographic and Atmospheric Administration, Earle advocates for ocean conservation and education.

While Musk explores the distant stars through SpaceX, Earle explores the depths of the oceans. These two mindsets are both common among climate believers. I refer to them as Wizards and Prophets, based upon the descriptor coined by the acclaimed researcher and author, Charles C. Mann.

For nearly twenty years, Mann, a successful writer for *The Atlantic*, had been troubled by a realization that struck him just after his daughter was born. He was walking alone outside the hospital having just witnessed the miracle of human birth, when the paralyzing thought struck him, "When my daughter reaches my age, there would be ten billion people living on the planet. How does that even work?"

Years later, after countless interviews with scientists and experts, Mann gave birth to the answer in a book he titled, *The Wizard and the Prophet: Two Groundbreaking Scientists and Their Conflicting Visions of the Future of Our Planet.*

In Mann's book, the clash between two competing ideologies is epitomized by two environmental champions living in America at the beginning of the twentieth century. They were barely acquainted and had little regard for each other's work. But the two men were largely responsible for the intellectual blueprints that we use today for understanding our carbon and climate dilemmas. Both men thought of themselves as using new scientific knowledge to face a planetary crisis. But that is where the similarity ends.

For Norman Borlaug, the Wizard, human ingenuity was the solution to our agricultural problems. For example, by using the advanced methods of the Green Revolution to increase per-acre yields, he argued, farmers would not have to plant as many acres, an idea researchers now call the Borlaug hypothesis.

The views of William Vogt, the Prophet, were quite the opposite: He believed the path to agricultural progress was to use ecological knowledge to get smaller. Rather than grow more grain to produce more meat, Vogt argued that humans should "eat lower on the food chain" to lighten the burden on Earth's ecosystems. Vogt's counter to Borlaug's hypothesis: farmers may grow more food per acre, but at the cost of wrecking the world's ecosystems.

Two Mindsets Emerge

Norman Borlaug was born in 1914 and raised in rural Iowa. An agronomist, he was at the forefront of "techno-optimism"—the popular belief, which continues today, that science and technology, when properly applied, will allow humans to innovate our way out of our problems. Borlaug, considered the father of the Green Revolution, won the Nobel Peace Prize in 1970 for pioneering agronomic techniques that increased grain harvests around the world, saving tens of millions of lives from starvation. Borlaug evangelized that wealth and affluence were not the problems but the solutions. Only through affluence and more knowledge could humans create the science that will resolve our environmental dilemmas: *Innovate! Innovate! Innovate!* was his gospel.

William Vogt was born in 1902 and raised on old Long Island, New York. An ecologist, Vogt laid out the basic foundations for the modern environmental movement. He advocated "apocalyptic environmentalism"—the belief that unless humans drastically reduce consumption and limit population, they will ravage global ecosystems. Through best-selling books and dynamic speeches, Vogt evangelized that wealth and affluence are not our greatest achievements but our biggest problems. If we keep taking more than the Earth can give, he said, the inevitable result will be devastation on a global scale: *Conserve! Conserve! Conserve!* was his gospel.

What Mann has written is more than just an examination of two men and their philosophies about human impacts on the earth—it is a lens through which we can see ourselves. If you believe that humans should use science and technology to innovate and improve life on Earth, then you are likely a Wizard. If you believe that humans and the earth have natural limits and we transgress these limits at our peril, then you are likely a Prophet. But what if both mindsets appeal to you?

The viewpoints of Wizards and Prophets are often described as opposing each other. One says, "Be smart, make more, and that way everyone can win." The other says, "Hunker down, conserve, obey the rules, otherwise everyone is going to lose." But is our world really so binary? Is it really black or white? Probably not.

Many of you will find yourself somewhere in the middle. In fact, most people describe themselves as a hybrid of the Wizard and Prophet. How is that possible?

Let's say you view climate mostly through a technology lens—you want to buy an electric car or solar panels for your home. This appears to make you a Wizard. Yet you also believe in protecting public lands and limiting private development in open spaces.

That's definitely a Prophet. In this case, you might *lean* more Wizard than Prophet.

Or perhaps you view climate mostly through a conservation lens—you are vegan and don't own a personal vehicle. This seems to make you a Prophet. Yet you also believe in decarbonizing the electric grid and electrifying transportation. Well, that's surely a Wizard. In this scenario, you probably *lean* more Prophet than Wizard.

In our opening story, you may not have guessed that Elon Musk is a *hybrid*—both Wizard and Prophet. His bold notion to power the entire US electric grid with one large solar farm (with battery storage) is totally Wizard. But his belief that it can be done with a limited footprint, focusing on land conservation, reveals a bona fide Prophet.

In reality, most of us fall somewhere in the middle of identifying as a Wizard or Prophet. *What's most important is that you know your leanings.* Take some time to reflect on how you think, talk, and act on carbon and climate, and you'll discover which mindset or balance of mindsets identifies you.

Acting On Your Mindset

Now that you have a sense of your leanings toward a Wizard or Prophet mindset, let's examine how this understanding can help you achieve your carbon and climate goals.

We opened this chapter by comparing two very different believer mindsets—Elon Musk and Sylvia Earle. Musk is one of the greatest technology entrepreneurs on the planet who leans mostly toward a Wizard, while Dr. Earle is one of the greatest marine biologists who has ever lived and who appears to lean mostly toward a Prophet. I was trained as a chemist and have been a clean-tech entrepreneur for decades. I am no Elon Musk or Sylvia Earle; but should I feel any less capable of acting on my Wizard or Prophet goals because of that? Absolutely not, and neither should you.

Remember, Elon Musk didn't invent electric cars or solar panels, but he took a business philosophy and vision and acted upon them to create companies that reflect his Wizard mindset. You and I can also act on our Wizard mindset through the beliefs we hold and the products and services we choose. Personally, I have become an evangelist for electrification. I recently purchased and installed a smart thermostat in my home. Within the next twelve months, I plan to buy an electric vehicle. Finally, within the next twenty-four to thirty-six months, I plan to install solar panels and battery storage at my home. These are just a few of the commitments that reflect my Wizard mindset. I do not have the resources of a billionaire, but that doesn't make my efforts any less significant, and neither are yours.

Likewise, Sylvia Earle did not create the science of marine biology, but she had the talent, intelligence, and courage to overcome barriers and ignorance of those that didn't feel she belonged. She is considered a Living Legend by the US Library of Congress because she followed her calling to the Prophet mindset. You and I can also act on our Prophet mindset by the values we embrace and decisions we make every day of our lives. For example, I have resolved to do no harm to others as my prime directive on carbon and climate, like my COVID-19 decision to wear a mask in certain settings to protect the health of others. I have resolved to eat 100 percent plant-based breakfast on weekdays. Finally, I wrote this book to help others discover the beliefs, values, and practices that can heal carbon addictions and lead to a sustainable life.

Focus on achieving your personal goals and remaining true to yourself; those are the true measures of success. Everyone has different assets to draw from in choosing their path of action—financial resources, education, information, and connections to other people. The suggestions I have shared are my personal choices, and they may or may not be right for you. Ultimately, only you can know which aspects of the Wizard and Prophet mindsets calls to you, and how you can use your personal resources to live your beliefs.

Reflections: Mindset

This chapter examined the complimentary nature of the Wizard and Prophet mindsets and the value of both innovation and conservation in climate change. Use the innovation lens to express optimism in benevolent use of data and technology to arrest climate change and decarbonize human civilization. Use the conservation lens to promote carbon minimalism—less is more—and embrace your shared, multigenerational responsibility to steward the earth.

While many people will thrill at the Wizards' engineering advances and put faith in the Prophets' cautious forecasts, they will also discover that technological miracles can produce nasty side effects and sacrifice often proves contrary to human nature. Think for yourself. Use the mindset creed to determine your own natural leanings and path to a sustainable life.

MINDSET CREED

I believe a sustainable human society
requires both innovation and conservation.

Benchmark

It is not enough to know your carbon mindset; you must act on that knowledge by setting personal climate change goals.

Which of the two mindsets calls to you? Resolve to take action: (1) read The Wizard and the Prophet by Charles C. Mann in the next six months, and (2) set three short-term (three, six, or twelve month) climate goals to reduce your carbon footprint, based on the mindset creed.

CHAPTER 4

SCIENCE

"The saddest aspect of life right now is that science gathers knowledge faster than society gathers wisdom."
—ISAAC ASIMOV

"If control is the problem, then, by the logic of the Anthropocene, still more control must be the solution."
—ELIZABETH KOLBERT

This book is written with the assumption that you believe the science behind climate change. The science tells us the earth is warming at an unprecedented level. Studies show that carbon dioxide concentrations are currently higher than they've been at any point in human history, global temperatures are rising at unprecedented rates, and warming is poised to surpass anything the planet has experienced in millions of years.

A recent study led by Raphael Neukom of the University of Bern investigated the scope of natural warming and cooling events experienced by the planet since the start of the Common Era. The data imply that it's not just unusual that the temperatures themselves are high; the fact that temperatures are rising all over the world has seemingly never happened before, at least not during the past 2,000 years.

Historic data show that human activity began to accelerate global temperatures during the Industrial Age. Starting in mid-eighteenth-century Great Britain, the Industrial Revolution ushered in an era of great prosperity for some people, defined by mass production, advanced agriculture, the rise of the nation-state, electric power, modern medicine, and running water. The quality of human life has increased dramatically during the Industrial Age, all the way until the present. Advances in sanitation, food production, safety, and health care mean that life expectancy worldwide today is more than twice as high as it was when the Industrial Age began.

However, the consequence of this revolution has been unfettered growth of greenhouse gas emissions, specifically carbon dioxide (CO_2), which we shorthand as "carbon" in this book. Carbon emissions are the main driver behind the greenhouse effect: Some gases in Earth's atmosphere act like the glass in a greenhouse, trapping the sun's heat and stopping it from leaking back into space, and thus causing global warming. Many of these greenhouse gases occur naturally, but human activity is increasing concentrations of some of them in the atmosphere. Carbon dioxide produced by human activities is by far the largest contributor to global warming. By 2020, its concentration in the atmosphere had risen to 48 percent above its pre-industrial level (before 1750).

During the nineteenth and twentieth centuries, scientists began to measure and understand the harmful implications of rising carbon dioxide levels to the climate. In 1896, a seminal paper published by Swedish scientist Svante Arrhenius first predicted that changes in atmospheric carbon dioxide levels could substantially alter the surface temperature of the planet through the greenhouse effect. Despite this early warning, world leaders chose the immediate benefits that came with the Industrial Revolution over a sustainable economic path that would benefit people and the planet. This has led to a zero-sum mindset when it comes to climate change. In a zero-sum scenario, people believe that whatever is gained by one side is lost by the other. Supporters of this mindset will argue that if you want the benefit of a sustainable economy, your current standard of living must be reduced. This thinking is a false narrative because there are no opposing sides in climate change. We are all on the same side. We all benefit from a sustainable industrial economy. We all lose if climate change accelerates. Addressing carbon and climate change is not about reducing our standard of living; it's about improving our quality of life through sustainable living. *That* is what the science supports.

Humans and Earth: A Complicated Relationship

Humans are in control. We have been terraforming the earth at globally significant levels for at least 5,000 years. The word *terraforming* describes the process of deliberately modifying the atmosphere, temperature, surface, or ecology of a planet, moon, or other body to make it habitable by Earth-like life. For example, some people think humans will eventually *terraform* and colonize Mars to make it habitable by humans.

Yet for thousands of years human civilizations have been terraforming and changing our own planet to meet the agricultural needs of people. For example, 4,000 years ago in ancient Egypt, farmers engineered a system called basin irrigation, in which they constructed networks of earthen banks to form basins and dug channels to direct the Nile River floodwater into the basins, where it would sit for a month until the soil was saturated and ready for planting. Similarly, 5,000 years ago in ancient China, floods threatened to destroy rice crops that fed millions, so farming communities along the Yangtze River Delta built a series of high and low dams and levees, making it one of the world's largest and oldest hydraulic engineering systems ever designed. Groundbreaking as they were, these early feats of terraforming the planet were just the beginning. As engineering and technology have advanced, so has the human impact on the natural systems that once governed the planet itself.

Today we have reached a new apex of human impact on the planet. We live in the Anthropocene, a new geological age in which humans are the primary cause of planetary change. The word combines the root *anthropo*, meaning "human," with *cene*, the suffix for "epoch" in geologic time. The distinctive human signature of this age is the human species bending the earth to meet their needs and the unintended consequences of these actions—both helpful and harmful.

We witnessed some of those unintended consequences during the COVID-19 outbreak. The global pandemic revealed much about us—as individuals and societies. It laid bare our strengths and our weaknesses. The global economic shutdown, while not ideal or desired, gave us a planetary sabbath from human-generated carbon emissions. People switched from driving to work to telecommuting. We saw nature and wildlife emerge in spaces previously dominated by development and commerce. The surreal sightings of pumas strolling Santiago, Chile, wild boars trotting traffic-free streets in Haifa, Israel, and masses of pink-plumed flamingos in Mumbai, India's, waterways are all testimony to what scientists refer to as the anthropause—the slowdown in human activity caused by the pandemic.

This unexpected consequence of the pandemic should strengthen our belief that the physical harm caused by climate change can be reversed if we take the necessary, bold steps to decarbonize our economies.

In addition to the ecological dimensions of climate change, we should consider the psychological effects. Most people believe that human activities have caused the planet to warm or worsened the problem. Yet climate change often feels overwhelming and beyond human control. People feel helpless and defeated by the magnitude and scale of the problem. As people grapple with a warming world and increasingly unstable weather, their mental health may be at risk. Many scientists predict that stress, anxiety, depression, and posttraumatic stress disorder will increase as climate change's physical impacts accelerate.

The psychological effects of climate change on the young are especially acute. In a recent study published in *Nature*, world-class researchers interviewed 10,000 young people in ten countries to find out how they felt about climate change and their government's responses to it. The data reveal a growing "climate anxiety" among the respondents. Most young people ages sixteen to twenty-five said that climate change made them feel sad, afraid, and anxious. That shouldn't come as a surprise. Climate change is likely to color every part of young people's lives—from the jobs they hold to where they call home.

Psychologist Per Espen Stoknes has spent years studying the defenses we use to avoid thinking about the demise of our planet. He has identified five psychological barriers to climate action: *distance*, the issue seems remote; *doom*, if it is seen as a disaster that requires sacrifice, it creates a wish to avoid; *dissonance*, if what we know conflicts with what we do, then we will try to not know it; *denial*, if we can deny that it is happening we can continue to feel good about ourselves; and *identity*, "cultural identity overrides the facts."

Can you relate to any of these? I certainly can. My own mental barrier arose when I realized that I was driving one hundred miles roundtrip every day from my home in Virginia to Washington, DC, to meet with lawmakers about curbing carbon emissions. I first tried to dismiss my ICE vehicle emissions as a small sacrifice for the greater good, but the truth was I needed to make a change. I now take the train 90 percent of the time and enjoy a zero-stress commute and a cleaner conscious.

When we take a close look at these psychological barriers and apply them to ourselves, we find that the biggest obstacle to action is in our own head. As individuals and communities, we must change our climate narrative from despair to hope. We must move from paralysis to action. We must replace the barriers in our minds with beliefs, values, and practices built on the scientific foundation of decarbonization.

Science and Clean Technology

Let's look at how science can help us solve the problem of climate change. Science is concerned with accumulating and understanding observations of the physical world. Once we understand a problem, we can use that knowledge to develop solutions. When we use scientific knowledge to tackle the problem of climate change, solutions take the form of clean technology.

The Intergovernmental Panel on Climate Change (IPCC) projects that at our current trajectory, the atmosphere is likely to warm by 1.5 degrees Celsius by 2040. The IPCC has stated that even if humans significantly reduced emissions, climate change would continue due to the accumulated emissions in the atmosphere. So where does that leave us? Let's consider the innovative approach one well-known technology group is pursuing to address the carbon crisis.

Y Combinator (YC) is a Silicon Valley startup incubator that has launched over 3,000 companies including Airbnb, Coinbase, DoorDash, Dropbox, Reddit, Stripe, Scribd, Twitch, and Weebly. In 2018, YC announced a request for carbon-removal technology startups and ideas came flooding in. This call and the response to it reveal how YC views the current climate problem and the path to science-based solutions.

YC has been an active investor in clean energy for many years. YC calls deploying renewable technologies and reducing future emissions critical strategies—this is Plan A. YC has assessed the crisis and determined that these actions alone cannot reverse the harm already done to the climate. The biggest issue is the scale of the task. Globally we currently emit about forty billion metric tons of carbon dioxide per year. That's a hard number to visualize because it's so large and carbon is invisible to the naked human eye. YC offers a simple reference: one metric ton of CO2 in solid form, measured and stored at standard atmospheric pressure, would fill a cube roughly the size of a three-story building. Now imagine 40 billion three-story buildings worth of CO2 being released every year. That's how big the problem is.

YC believes that halting global climate change now requires both Plan A and a Plan B: removing carbon dioxide from the atmosphere in an affordable manner. According to IPCC analysis, the 1.5 degrees Celsius goal is unachievable without carbon dioxide removal. We already know how to remove carbon from the atmosphere. Unfortunately, methods like forest management, planting forests, and sequestering carbon in the soil are extremely land intensive.

While there are numerous ways to capture or remove carbon, two methods dominate the investment landscape: avoiding emissions from heavy industry and removing CO2 from the atmosphere. Carbon-capture technology aims to stop the CO2 emissions produced by power plants or industrial operations from entering the atmosphere. Carbon-removal technology pulls CO2 already present in the atmosphere out of the air. YC is focused on carbon removal.

While YC is interested in anything involving carbon removal, their announcement highlighted four approaches to carbon removal that they'd like to see explored more.

1) Ocean phytoplankton.

Genetically engineered phytoplankton might be used to photo-synthetically convert carbon dioxide into a carbon sequestration medium. In order to remove about forty-seven billion metric tons per year, about 1 percent of the ocean's surface would need to be assimilating CO2 at full rate.

2) Electrogeochemistry.

Mineral weathering is a geochemical process in which carbon dioxide reacts with common rocks, forming dissolved mineral bic-arbonates. Some one billion metric tons of CO2 per year, about one-fortieth of human annual CO2 emissions, are removed naturally from the atmosphere by this process. A range of ninety to nine hundred billion metric tons of CO2 per year could be removed using this method.

3) Cell-free enzyme systems.

Microbes can make enzymes, that can be leveraged to do many things, including synthesizing CO2 into other compounds. However, the consequences of this technology are not well known and could be more dangerous than climate change itself. YC predicts that this approach could remove up to ten billion metric tons of CO2 per year.

4) Desert flooding.

About 10 percent of the world's surface is desert, which is cheap, uninhabited, unproductive land. This terraforming approach seeks to leverage these attributes by flooding the desert with water to create millions of one square kilometer oases that can grow phytoplankton that will capture and sequester CO2.

YC acknowledges these are ideas that "straddle the border between very difficult to science fiction." However, if any of these innovations work, they could help solve our climate problems more affordably—for billions of dollars rather than tens of trillions of dollars. That sounds like a gamble worth taking.

Reflections: Science

This chapter has encouraged trust in scientific evidence that the earth's climate is warming and human activity is the primary cause. We live in the Anthropocene, a geological age when humans are the primary cause of planetary change. Humans have ambitions to become an interstellar species; however, we must first learn to live sustainably on the earth. That means using science-based targets and technologies to reduce and remove carbon emissions. It also means understanding the psychological effects of climate change on individuals and societies. We must reject the false zero-sum narrative of deniers of climate science. Use the science lens to strengthen your belief that rising carbon emissions present an existential threat to human health, the economy, and the environment. Each of us has a role in reducing the human carbon footprint, no matter how small.

SCIENCE CREED

I believe in the scientific consensus that the earth's climate is warming and human activities are the primary cause.

Benchmark

During the late nineteenth century, scientists began to measure and understand the harmful implications of rising carbon dioxide levels on Earth. Despite those early warnings, world leaders chose the immediate benefits of the Industrial Revolution over a sustainable economic path. Today science is again warning that human civilization must decarbonize to avoid a climate catastrophe.

What convinced you to believe the science of climate change? Be resolved to do these things: (1) read How the World Really Works by Vaclav Smil within the next six months, and (2) ask three young people (six to eighteen years old) if they believe that adults (over thirty years old) will solve climate change, and why?

CHAPTER 5

KINSHIP

---

*"Always remember that to argue, and win, is to break down the
reality of the person you are arguing against. It is painful to
lose your reality, so be kind, even if you are right."*
—HARUKI MURAKAMI

*"Never have human societies known so much about mitigating
the dangers faced from [climate change] but agreed so little
about what they collectively know."*
—DAN KAHAN

The carbon and climate change story is one that humans
have struggled to tell convincingly for half a century.
The problem is surely not a lack of facts and figures.
The best scientific minds on the planet have routinely
produced one epic report after another detailing the
adverse impact humans have on Earth since the dawn
of the Industrial Revolution—to little avail.

The problem can't be blamed on delivery either. The new millennium has brought forth an explosion of methods for communicating about carbon and climate, spanning nearly every conceivable medium and messenger. Most of these are attractively designed and well executed, careful to avoid shortcomings that have stumbled previous efforts. Still they struggle to gain traction.

If it's not data or delivery, the problems the climate narrative faces must run much deeper: the nature of the issue itself.

Climate change is abstract, uncertain, impersonal, and can feel distant despite the growing scale and frequency of climate-related crises across the planet. Even if there were no explicit efforts to confuse and divide the public, the climate narrative would still be challenging to talk about in ways that motivate public engagement and mobilization rather than inspire despondency and resignation.

The tragic irony is that the story of carbon and climate is deeply human—we caused it, we will suffer from it, and we alone can take the necessary action to avoid its worst consequences.

Yet translating climate change from a scientific reality to a societal imperative has proven extremely difficult. Many people still consider it an "environmental" issue. I believe the most effective long-term approach to getting diverse audiences to engage deeply with climate change requires that we stop treating it as a stand-alone problem. We need to fundamentally rethink and alter the way we talk about and position climate change. Perhaps the best strategy is to say that climate change is a health risk and a risk to humanity's survival. That's the approach we have taken here—that climate change is the story of humans as the dominant species on the planet. Throughout this book we link carbon addiction and the climate crisis to human health. Each chapter presents an aspect of the creed designed to make it easy to self-diagnose and prescribe an action plan that works for individuals.

Far too often climate communication is presented as hard facts and figures. But how important are facts and accuracy to the human mind? Probably not as much as you think. Especially when it comes to carbon and climate.

The unfortunate reality is that the words "climate change" have been weaponized. This serves as a distraction from the originally intended message. Many people view climate change as code word in a zero-sum ideological narrative, as discussed in the science creed chapter. People hear those words and automatically take sides, either for or against immediate decarbonization. This failure to effectively communicate the urgency about carbon and climate has divided humans and created a conflict between different groups' convictions and knowledge.

The default to tribalism, dividing over different beliefs or knowledge, is not surprising. Throughout human history, our ancestors have lived in tribes. Becoming separated or cast out from the tribe was like a death sentence. Consider the story of Giordano Bruno: He was burned at the stake in 1600 in Italy by his own tribe (the Catholic church) for his beliefs, including that the earth orbited the sun, that the stars were distant suns surrounded by their own planets, and that these planets might foster life of their own. It did not matter to anyone other than his tribe, of course, whether this was true or not, and they killed him for his beliefs (which mostly turned out to be true).

In recent years, the word *tribe* has taken on a broader meaning. Today it is frequently used to refer to any social group with a strong identity or affinity (e.g., national, ethnic, or political). Increasingly, this social affinity can warp a person's ability to reason, resulting in tribal thinking, which trumps any notion of truth or scientific evidence to the contrary.

Tribal thinking may seem irrational, but there is a perfectly rational reason people put tribal loyalty ahead of truth. Most false beliefs do not result in any direct harm to the individual holding them. It makes no difference whatsoever to an individual whether they believe in human-induced climate change or not, but it makes a ginormous difference if that belief goes against the dominant thinking of their tribe. That person may be seen as disloyal or even as a traitor. In the past, going against the beliefs of the tribe may have put your life in danger (e.g., Giordano Bruno), and today you could easily be expelled from your social circle, lose your job, or be ostracized by family members.

Believing the truth is important, but so is remaining part of a tribe. While truth and tribe often work well together, when we must choose between the two, people often select friends and family over facts. This explains why you might bite your tongue when a climate denier hijacks the family Thanksgiving dinner or wince when your best friend says something ignorant about climate science. In both cases, the tribe trumps truth and facts.

When it comes to effective communications, I believe we should learn from those who do it well. One of my favorite writers and communicators is self-help guru James Clear, author of the best-selling book *Atomic Habits*. Clear writes a weekly newsletter titled 3-2-1, which he describes as "The most wisdom per word of any newsletter on the web." I agree with his description and think you will too once you read his work.

In 2019, Clear wrote a whitepaper titled "Why Facts Don't Change Our Minds," which is particularly relevant to our discussion about tribes and climate. In the paper, he observes that "humans are herd animals." I agree with this assessment. One of our deepest desires it to belong. We want to fit in, bond with others, and earn the respect and approval of our peers. These inclinations are essential to our survival and help explain our tribal tendencies.

Understanding the truth of a matter is important but so is being part of a tribe. In most situations, social connection is more valuable to your life and happiness than understanding a particular fact. This is where scientific climate facts and figures can fail.

Clear then hits upon the crux of the matter. "Convincing someone to change their mind is really the process of convincing them to change their tribe." People will not change their mind if you take away their community and leave them hanging. You must give them somewhere inviting to go, somewhere they will feel welcomed not judged. Clear suggests that we bring them into our fold. To successfully change people's minds about carbon and climate you must become their friends and bring them into your tribe. That allows them to change their beliefs without the risk of being socially isolated or abandoned.

This leads us to the kinship creed. In his paper, Clear notes that the word kind originated from the word *kin*. He proposes that being *kind* to someone means you are treating them like family. This framing can be immensely helpful to climate communications. The kinship creed says, "When seeking common ground about carbon and climate change, it is more effective to ask and listen before speaking." As an example, during a meal with extended family, it would be hospitable for the host to invite the guests to partake of the meal first. In a similar way, we should welcome and listen to doubters before sharing our thoughts. Taking this kinder, more thoughtful approach to carbon communications shows a willingness to hear other views and seek common ground, and it tends to yield positive results.

The Lost Tribes

Now that we have a plan about how to communicate about carbon and climate, we need to identify our target audience. Most Americans accept the science creed which states, "The earth's climate is warming, and human activities are the primary cause." National surveys report that 60 percent of the country thinks this way. That makes the majority of Americans climate believers. The problem is only 30 percent of Americans talk openly about climate change.

This silence among climate believers is hampering efforts to make climate change a clear, national priority. By remaining silent, the believer majority is giving the denier minority an outsized voice in the climate narrative. To change this dynamic, the believers need to have more open conversations about carbon by harnessing hope, trust, and shared values. I recommend focusing on clean technologies like electric vehicles, solar energy, and battery storage, all of which enjoy overwhelming public support. Effective communication on climate starts with addressing the likes and needs of your target audience. Remember they are kin.

Now, let's review some facts and figures to help identify our target audiences. Researchers at Yale and George Mason University annually break down the spectrum of US perspectives on humanity's role in climate change into Six Americas: (1) *dismissive,* 9 percent; (2) *doubtful,* 9 percent; (3) *disengaged,* 5 percent; (4) *cautious,* 17 percent; (5) *concerned,* 30 percent; and (6) *alarmed,* 29 percent. The largest combined share of the public—60 percent—are concerned or alarmed.

The *alarmed* are those most certain that climate change is happening, most concerned about it, and most supportive of climate policies and actions. The *concerned* also think climate change is happening and view it as a serious threat, but they tend to perceive it as more distant and less urgent than the alarmed. The *cautious, disengaged,* and *doubtful* are each at different stages of understanding climate change and are less engaged on the issue. The *dismissive* reject the reality and threat of human-induced climate change and oppose taking action to address it.

These data show that by focusing on the cautious segment (17 percent), total believers could push public opinion beyond the tipping point of 70 percent in favor of climate action. By converting the cautious to believers, the prevalence of the zero-sum narrative of climate winners and losers would be diminished. The partisan divide on carbon would be weakened, if not removed. This is the first target audience.

The research professors at Yale and George Mason have also identified a second target audience by examining the distribution the Six Americas across the three largest racial and ethnic groups in the US: non-Hispanic/Latino Whites, non-Hispanic/Latino African Americans, and Hispanics/Latinos. The data show that Americans across all three racial and ethnic demographic groups are alarmed or concerned about climate change, but Hispanic/Latinos (69 percent) and African Americans (57 percent) are more likely to be alarmed or concerned about climate than whites (49 percent). In contrast, whites are more likely to be doubtful or dismissive (27 percent) than are Hispanic/Latinos (11 percent) or African Americans (12 percent). This gives believers two additional demographics to target. As the US becomes more racially and ethnically diverse, promoting public engagement on climate change across diverse audiences is becoming increasingly important.

Reflections: Kinship

This chapter has shown that facts don't change minds, people do. Convincing someone to change their mind is really the process of convincing them to change their tribe. You can use this lens to discuss climate with people who do not necessarily share your convictions on carbon. You can put the kinship creed into practice by evangelizing about decarbonization in word and deed. Learn to listen, be kind, and find common ground. Remember, all humans are kin.

KINSHIP CREED

When seeking common ground about carbon and climate change, I believe it is more effective to ask and listen before speaking.

Benchmark

The story of carbon and climate change is a human story—we caused it, we suffer from it, and we must take the urgent action required to avoid its worst outcomes. We must overcome our tribal tendencies with kindness.

Which has a greater influence on your climate beliefs—facts or friendships? Be resolved to do the following: (1) verbally share your carbon beliefs with at least four people (in word), and (2) share this book with at least four people (in deed).

CHAPTER 6

HABITS

"We are what we repeatedly do. Excellence, then, is not an act, but a habit."
—WILL DURANT

"All we have to do is to wake up and change."
—GRETA THUNBERG

Ultimately, our lives are not shaped by big decisions or events but by our daily habits. What we repeatedly do defines who we are. In fact, some habits can be elevated and given greater meaning. In this book, when a habit becomes an intentional practice that has meaning, we call it a ritual.

Rituals can transform our lives by turning small, everyday actions into character-defining practices. Unlike routine habits, such as brushing your teeth, rituals embody what matters to you and put your values into practice. Running out of the door with a bagel in hand before work is a habit, but sitting down and saying grace over coffee and a bagel before heading to work is a ritual.

Rituals engage your heart and mind—they build character and give you greater focus. Consider the rituals practiced by the great American poet Maya Angelou. To optimize her creativity and mental focus, Angelou would rent a hotel room and remove all possible distractions such as photographs, books, or TV. She started working every day at 7 a.m. sharp. Armed with a bottle of sherry, a deck of cards, legal pads, a thesaurus, and the Bible, she spent hours writing in this purposely designed environment.

Anthropologists have found that humans turn to rituals when faced with a situation where the outcome is *important, uncertain, or beyond our control.* Climate change checks all three boxes. Building rituals can help you combat climate anxiety by turning small habits with little immediate impact into meaningful practices that compound to make lasting change and impact in our lives and on Earth. Equipped with this knowledge, you can use rituals to embrace the Carbon Creed in your life.

In this chapter, I present an unorthodox approach to rituals that I believe will help you on your decarbonization journey. The inspiration comes from the book *The Power of Ritual* by Casper ter Kuile. In his book, Casper uses ecclesiastical terms to explain how he has turned everyday activities into soulful practices. My hope is that we can help you transform everyday habits into carbon-conscious practices that have greater meaning and purpose. With that in mind, I am pleased to share the carbon-conscious rituals: sacred text, sabbath rest, sanctuary, scribe, and supper.

SACRED TEXT

Read, listen, and think deeply about something that inspires you.

In this book, the ritual of reading makes a text sacred, not the text itself. Thus, anything you read that inspires you on your carbon-conscious journey can become a sacred text. You can convert any text into a reading ritual. Whether it's a daily devotional or bedtime story with a child, you could practice a reading ritual every day. Let's consider a few examples.

One popular reading ritual is consuming the daily newspaper. I always start my day by reading *The New York Times*, *The Wall Street Journal*, and my local paper on my iPad. You can make this morning ritual carbon conscious by (1) changing your subscriptions from paper to electronic, (2) reading with an eye to carbon and climate stories and sharing them on social media, or (3) gifting someone an electronic subscription.

Another common reading ritual is joining a book club. There are several celebrity book clubs that you can join (including Oprah Winfrey and Reese Witherspoon), but why not start your own? You can make your book club participation carbon conscious by (1) selecting climate-themed books, (2) choosing audiobooks or e-books rather than a paper, (3) sharing your personal copy with someone, or (4) reading a book as a group.

Reading is one of the most important life skills we learn. Use the gift of literacy to become more carbon conscious. (A list of book recommendations with carbon and climate themes is provided in the Resources.)

SABBATH REST

Pause to refresh, reflect, and focus your mind on the present.

In this book, sabbath rest means pausing or stopping a certain practice for a predetermined length of time. The practice of sabbath rest can refresh your body and mind. It can help you refocus on the present. Applying a carbon-conscious lens to your sabbath rest will enhance the ritual in many ways. Let's consider a few real-life examples.

Today our lives are ruled by technology. Adults spend upwards of eleven hours every day on screens—

most of which are for personal use, not work. That means the average person likely spends more time on their screens than sleeping!

It's impossible to imagine a world without mobile phones, but it's easy to overlook the impact that they have on our climate. Using your phone each day has a bigger carbon footprint than making it. A lot of energy is required to run the data centers and infrastructure that allow you to make Zoom calls, post selfies, and stream videos every day. In fact, some of the world's largest data centers require more than one hundred megawatts of power—enough to power around 80,000 US households. This is why you need a carbon-conscious phone sabbath.

Practicing a phone sabbath will not be easy. I know this firsthand because I practice a phone sabbath every week. I recommend that you start with a twelve-hour sabbath. Twenty-four hours of "silence" might be too long for friends, family, and colleagues.

I also suggest you pick a time frame that includes your sleeping hours so that if you use your phone for work, it won't disrupt your professional life. My original phone sabbath started at 7 p.m. on Friday night and lasted until 7 a.m. on Saturday morning. That avoided any interruption of my work schedule and allowed me to have a Friday date night with my wife.

Over time, you may find it easier to add more devices to the sabbath than to lengthen the time of your sabbath. Today, my twelve-hour sabbath has evolved to include my phone, tablet, and laptop. I now call it a *tech sabbath*. You can also gradually lengthen the time of your sabbath, perhaps adding four-hour increments, until you reach a goal of twenty-four hours. I now practice a twelve-hour sabbath three nights per week.

These are just a few of the countless ways you can apply the sabbath principle to everyday activities (e.g., sabbath from driving). The key is to identify the habits in your life that could benefit from a sabbath rest and make them carbon-conscious rituals.

SANCTUARY

Find a safe, serene space or environment that relieves climate anxiety.

During the COVID-19 pandemic, our homes became our fortresses. It's where people worked, taught their children, and attended church. Even in less urgent times, home is where people sleep, eat, and entertain. So it's helpful to make your home into a place you really want to be. It should feel like a sanctuary. Distress over the climate crisis can affect your mental health and well-being. To help you cope with climate anxiety, make your home a carbon-conscious sanctuary.

Creating a spot in your home solely for relaxation will help you to cultivate a *ritual* of relaxation. To help you focus, remove all distractions including screens, electronics, pets, or people. This space can be anywhere—your spare bedroom, basement, attic, closet, balcony, screened-in porch, or shed. To make your sanctuary carbon conscious, incorporate energy efficient features such as LED lighting, a smart thermostat, and smart electrical outlets.

Channel the atmospheres of your favorite places into your sanctuary. Reflect on how your favorite café or wellness spa evokes a sense of peace. Maybe the café has comfortable seating and the scent of coffee. Perhaps the wellness spa diffuses essential oils and has a minimalist feel. To add ambiance to your sanctuary in the early morning and evening, try lighting beeswax or soy candles with cotton or paper wicks, which are mostly carbon neutral.

As you plan your home space, think about what types of sensory triggers allow you to relax. You might play jazz or classical music while cooking, arrange potted plants throughout your home, bake chocolate chip cookies, or put wind chimes on your patio. When it comes to music, a few of my favorite carbon- and climate-themed albums include Marvin Gaye's *What's Going On* and Sting's *The Dream of the Blue Turtles*. There are countless other artists and genres that you can add generously to your own carbon-conscious playlist.

Creating a safe, serene space that becomes your sanctuary, will enhance your desire to practice carbon rituals in your home.

SCRIBE

Write a journal, penning words to express your thoughts and feelings about climate.

The practice of scribing is a powerful self-care ritual. It helps you capture your authentic self in words.

The word *scribe* comes from the Latin root word *scrib* (or *script*), which literally means "write." In this book, we use the verbs *scribe* and *journal* interchangeably. The act of putting thoughts into written words can help you articulate your feelings, thus transforming your feelings into something real, tangible, and manageable. It allows you to really listen to yourself as you might listen to someone you care about.

By keeping a journal, you are taking control of your thoughts. What you write and what happens to that writing is your decision. You can even discard your words after writing them if you like, but the act of writing is important and transformative. When journaling your thoughts on carbon and climate, there can be so many different aspects to the problem that it can feel completely overwhelming. Writing about your day-to-day life, reflecting on your own relationship with the planet, and documenting the carbon rituals you practice is a very helpful way of coping with stress and anxiety.

Most importantly, scribing with purpose allows you to deconstruct the climate crisis while staying optimistic about the future, helping you to believe in the possibility of positive change.

SUPPER

Share a meaningful connection over an informal meal with friends.

There is something incredibly special about being invited into someone's home for a meal. We get a glimpse into their daily lives and gain a deeper understanding of who they are by the ways they organize their books, arrange their furniture, position the plants, and decorate the kitchen. It's intimate.

That's why you should start a supper ritual with friends. For most people, supper is the main social meal of the day. I love this ritual because I'm a social person and I know a tried-and-true format that works well. I'm happy to share it with you here.

First, invite a group of folks (two to eight people is ideal) to your home for a potluck supper. Suggest a cultural theme (e.g., Italian, Southern, Asian, Mexican, Caribbean), but be sure to encourage carbon-conscious suggestions (e.g., chicken and fish over beef and lamb to reduce the carbon footprint of the meal and to make it healthier).

At dinner, invite everyone to share a short summary about the changes in their lives since you last came together (or since a milestone event). After dinner, invite the group to a common space where everyone can sit and relax comfortably. Light a candle to add ambiance. Then, one person at a time, share something you are excited by and something you are challenged by. Try to make one of them carbon conscious. Everyone listens. No comments, no thanking, no critiques, no problem-solving. After that person's sharing, everyone shares a commendation for this person. Express what you see in them, what qualities you admire, or how you've seen them grow. Once everyone has shared, thank them all, play some cool music, and round it out with a dessert selection (beverage, food, or both).

Start your supper ritual today. Begin by planning just one. People will ask for more. In no time at all, you'll have an incredible community of friends with whom you feel more intimately connected and able to share your carbon beliefs and values.

In conclusion, I hope these carbon-conscious rituals will help you set and achieve your personal carbon goals. Many people are told that the actions they take to reduce their personal carbon footprint will have little to no impact on the climate. We must reject that view. When we take an individual action, the initial effect may be small, but the direct effect of what we do is not the end of the story. Your individual action can influence the thinking and decisions of others. You can trigger network effects. If you choose to buy an electric vehicle, maybe your friend will too. If you seek to telecommute, maybe your colleagues will too, causing your employer to rethink remote work policies. Be the change you want to see. Become a carbon influencer.

Reflections: Habits

This chapter has shown how to elevate everyday habits and routines into meaningful rituals that decarbonize your life and engage others. You can put this creed into practice by adopting the five carbon-conscious rituals: sacred text, sabbath rest, sanctuary, scribe, and supper. Remember, it is the ritual of reading that makes a text sacred, not the text. Take a sabbath rest to pause, refresh, and refocus your mind and reduce your carbon footprint. Create carbon-conscious sanctuary spaces and rituals at home. Scribe daily about your relationship with the planet and carbon rituals you practice. Host a carbon-conscious supper with friends.

While this template has proven effective for many, ultimately you are a unique individual with your own habits and carbon goals. There's not a right or wrong set of carbon rituals to follow. The secret to success is identifying the ones that inspire and motivate you to action and make you more carbon conscious.

HABITS CREED

I believe transforming everyday habits and routines into carbon-conscious practices gives more meaning and purpose to life.

Benchmark

What we do repeatedly defines who we are. Be the change you want to see. Adopt the carbon-conscious rituals sacred text, sabbath rest, sanctuary, scribe, and supper.

Which of the carbon-conscious rituals do you feel most drawn to? Rank them in order of importance to you today. Be resolved to begin these rituals: (1) start a daily reading ritual with a focus on carbon and climate themes (newspapers, books, blogs, newsletters), and (2) begin a weekly sabbath rest to reduce your carbon footprint (phone, driving, diet).

CHAPTER 7

ACCOUNTABILITY

"No one comes from the earth like grass. We come like trees. We all have roots."
—MAYA ANGELOU

"Humanity has to act globally, at speed and at scale, to meet the staggering challenge of decarbonizing the global economy."
—JOHN DOERR

Thus far this book has focused on personal beliefs, values, and practices of individuals on the path to decarbonization. Each of us needs to become carbon conscious and reduce our carbon footprint where possible. However, the climate problem cannot be solved by individual consumption choices alone. Governments and corporations created this problem, and both must be held accountable for a solution.

Climate change is a systemic problem, fueled by eight billion human souls all engaged in the global hydrocarbon economy. It is a planetary-scale threat and as such, requires planetary-scale action. To be effective, the international community and companies must enact global carbon agreements and pledges. Governments have the power to implement policies that compel corporations *and* individuals to embrace decarbonization.

In this chapter, we will focus on the growing movement by corporations and governments to adopt carbon pledges, what that means for the climate, and how you as an individual can keep them honest. This is the accountability creed.

The Corporate Pledge

It's hard to read the news these days without coming across a company promising to "reach net-zero" or go "carbon neutral" or even become "carbon negative." The common person can barely make sense of whether these pledges are truly game-changing or greenwashing.

A vast array of corporations, including Amazon, British Petroleum, Delta Airlines, Google, Microsoft, Salesforce, and Starbucks, have pledged to drastically reduce their carbon footprints by committing to environmental, social, and governance (ESG) goals.

But what do all the different pledges mean and how can you understand them? First, let's define the word *pledge*. A pledge is a solemn promise or agreement to do or refrain from doing something. A carbon pledge is often an aspirational goal for the distant future. That doesn't mean it's not important—at the very least, a carbon pledge signals that a company believes science—for many, that's a big deal. In the context of ESG, most companies typically take one of the following carbon pledges:

1) CARBON NEUTRAL. Carbon neutral means that any carbon dioxide released into the atmosphere from the company's activities is balanced by an equivalent amount being removed. This can include buying carbon offsets or credits. *Net-zero* is essentially the same as carbon neutral.

2) CARBON NEGATIVE. Carbon negative requires a company to remove more carbon dioxide from the atmosphere than it emits.

Now let's consider a few well-known corporate examples of each type of carbon pledge: *Apple* has pledged to become carbon neutral across its entire business, including manufacturing supply chain and product life cycle, by 2030. The company is already carbon neutral in its global corporate operations, and this additional commitment means that by 2030, every Apple device sold will have net-zero climate impact. *Amazon* cofounded The Climate Pledge—a commitment to net-zero carbon emissions across its business groups by 2040 (ten years ahead of the Paris Agreement). To date, fifty-three companies across eighteen industries and twelve countries have become signatories. *Microsoft* has pledged to become carbon negative by 2030—taking more carbon out of the air than what's produced by its operations and supply chain.

To understand the true level of ambition behind a carbon pledge, you must consider the company's size, sector, location, and any prior climate commitments. According to John Sottong, an economist at the World Resources Institute who helps manage its Science Based Targets Initiative (SBTi), asking the following key questions will give you a better sense of the strength of the pledge:

1. WHAT'S THE DEADLINE TO REDUCE EMISSIONS?

Preferably, companies should provide at least two deadlines: a near-term goal and a long-term one. Long-term goals should aspire to reach net-zero emissions no later than 2050. Short-term goals should be set for 2035 or earlier and serve as a stepping stone to net-zero. Really, anything beyond fifteen years is probably aspirational.

2. WHAT TYPES OF EMISSIONS AND TARGETS ARE INCLUDED IN THE PLEDGE?

Most carbon emissions can be divided into three scopes: Scope 1 covers direct emissions from sources the company owns and controls source (e.g., manufacturing); scope 2 covers indirect emissions from the generation of purchased energy (e.g., utility supplied electricity); and scope 3 covers indirect emissions linked to a company's value chain or operations (e.g., suppliers and products).

Embedded within each pledge is the carbon target. Most carbon targets can be divided into three distinct groups: (1) **Absolute target** means the company plans to eliminate all or most of its emissions completely. This is the gold standard. (2) **Net target** means the company may still produce emissions but will offset them in some way. This is the silver standard. (3) **Intensity target** describes the emissions per unit of company product. An intensity target still allows for net positive emissions. This is the bronze standard.

3. ARE CARBON OFFSETS INCLUDED IN THE PLEDGE?

This is where things get interesting. Let's discuss the carbon offset business model for a moment. Carbon offsets allow a company to invest in some form of carbon dioxide removal (CDR)—such as tree planting or capturing methane emissions from cattle—and then subtract the equivalent emissions from the company's carbon footprint.

The carbon offsets business model is not perfect. An offset can never perfectly guarantee emissions reductions. But don't make the perfect the enemy of the good.

There's nothing wrong with a company buying carbon offsets from a reputable vendor. The problem occurs when a company uses CDR offsets over absolute emissions reductions to meet their entire climate pledge.

As more companies follow the CDR route, the total volume of offsets they rely on will quickly exceed the ability of the planet to provide them. There will not be enough land and trees to soak up growing carbon emissions. The best way to use offsets is to go beyond carbon neutral. If companies plan to use CDR offsets and want to build public trust in their carbon pledge, they need to provide greater transparency about (1) how the offsets will be achieved, (2) how emissions will be kept permanently out of the atmosphere, and (3) near-term targets.

This information shows why it's important to understand the meaning and metrics behind carbon pledges. Most people assume when a company announces a carbon pledge that it's in good faith, but that may not be the case. Some companies use their carbon pledge to continue polluting under the halo of good intentions. That's not acceptable. Individuals have a right to know the true carbon impacts associated with the products and services they buy and use. As carbon-conscious consumers, we have the buying power to demand greater carbon transparency.

The Government Pledge
PARIS AGREEMENT

As one of his first acts in the Oval Office on January 20, 2021, President Joe Biden made good on an early pledge by signing an executive order to have the United States rejoin the Paris climate agreement, the largest international effort to curb climate change.

The US played a large role in creating the 2015 agreement, which aims to avoid the most catastrophic climate change scenarios by keeping average global temperatures from rising no more than two degrees Celsius and preferably less than 1.5 degrees Celsius, compared to pre-industrial times, by 2100. Global temperatures have already increased by a little more than one degree Celsius.

The Paris Climate Agreement reached its five-year anniversary in 2020. How effective has it been? Are nations on track to meet their carbon pledges? It is probably best to describe the Paris Agreement as "a work in process." The agreement is an unusual hybrid of soaring ambitions and few enforcement mechanisms. Every country in the world signed onto a promise to take steps to keep global temperature increases "well below" 2°C by 2100. Doing so would require decarbonizing the energy and transportation sectors, halting the loss of forests, overhauling food production, and finding ways to suck greenhouse gases out of the atmosphere. Yet to meet the overall goal, countries were allowed to come up with their own goals and plans for how to accomplish them. Falling short comes with few concrete penalties. This penalty-free model is a flaw that needs to be corrected.

A Bloomberg Energy analysis of the 184 pledges for 2030 found that almost 75 percent were insufficient—they would not reach their stated goals. In fact, the world's first and fourth biggest emitters, China and India, are on track to have *higher* emissions in 2030. The United States is the second largest emitter, and its pledge is too low, based on its 2020 stated plan. Russia, the fifth largest emitter, has pledged to cut emissions 30 percent by 2030 from 1990 levels. (Russia's emissions in 2012 were nearly 70 percent below 1990 levels, so emissions could increase and they would still meet their pledge.) Finally, the European Union, the third largest emitter, pledged to reduce emissions by at least 40 percent by 2030 and is likely to reach a near 60 percent reduction because of a more aggressive plan.

The enormity of the Paris Agreement tasks is daunting. According to the United Nations Intergovernmental Panel on Climate Change, the goal is to cap global warming at 1.5 degrees Celsius compared to pre-industrial level, by cutting carbon dioxide emissions 45 percent from 2010 levels by 2030 and to net-zero by 2050. Doing so will require rapid, far-reaching transitions in energy production, agriculture, transportation, and building designs, along with carbon dioxide removal.

Over one hundred countries have pledged to become carbon neutral by midcentury. Still global emissions have continued to increase at a steady clip, interrupted only by the anthropause induced by COVID-19. If we were to focus on pre-pandemic trends, the world is on track to exhaust its carbon budget by 2035. Despite the urgent warnings from scientists, international climate negotiations have largely failed to match the level of conviction needed to meet the challenge. How can we change the current global trajectory of growing carbon emissions?

While government officials may be sincere when making carbon vows, they aren't always as faithful. That's because leadership can change frequently, and with changing government comes broken vows. Government is much better suited for creating the policy incentives necessary to decarbonize our economies. I believe there are three immediate policy imperatives that all sovereign governments must enact, and corporate sectors must embrace, if we want to prevent a worst-case climate scenario:

1. adopt a carbon fee and dividend

2. decarbonize the grid

3. electrify transportation

CARBON FEE AND DIVIDEND.

The time has come to adopt a carbon fee and dividend policy to reduce greenhouse gas emissions and address climate change. The policy imposes a carbon fee on the sale of hydrocarbon products and then distributes the revenue of this fee over the entire population (equally, on a per-person basis) as a monthly income or dividend payment. Experts describe the carbon fee and dividend policy as an alternative to complex regulatory approaches, cap and trade systems, or a stand-alone carbon tax. Since the adoption of carbon pricing programs in Canada and Switzerland, the concept has gained increased interest worldwide as an economy-wide and socially equitable approach to tackling climate change.

The carbon fee and dividend policy has also garnered strong support from economists. On January 16, 2019, *The Wall Street Journal* published, "The Economists' Statement on Carbon Dividends"—a joint statement by leading US economists promoting a carbon dividends framework for US climate policy. Signatories include 3,623 US economists, four former chairs of the Federal Reserve, twenty-eight Nobel laureate economists, fifteen former chairs of the Council of Economic Advisers, and two former secretaries of the US Department of Treasury. The statement was originally organized by the Climate Leadership Council and is recognized as the largest public statement of economists in history. Below is the original statement in its entirety:

ECONOMISTS STATEMENT ON CARBON DIVIDENDS

Global climate change is a serious problem calling for immediate national action. Guided by sound economic principles, we are united in the following policy recommendations.

i. A carbon tax offers the most cost-effective lever to reduce carbon emissions at the scale and speed that is necessary. By correcting a well-known market failure, a carbon tax will send a powerful price signal that harnesses the invisible hand of the marketplace to steer economic actors towards a low-carbon future.

ii. A carbon tax should increase every year until emissions reductions goals are met and be revenue neutral to avoid debates over the size of government. A consistently rising carbon price will encourage technological innovation and large-scale infrastructure development. It will also accelerate the diffusion of carbon-efficient goods and services.

iii.   A sufficiently robust and gradually rising carbon tax will replace the need for various carbon regulations that are less efficient. Substituting a price signal for cumbersome regulations will promote economic growth and provide the regulatory certainty companies need for long-term investment in clean-energy alternatives.

iv.   To prevent carbon leakage and to protect U.S. competitiveness, a border carbon adjustment system should be established. This system would enhance the competitiveness of American firms that are more energy-efficient than their global competitors. It would also create an incentive for other nations to adopt similar carbon pricing.

v.   To maximize the fairness and political viability of a rising carbon tax, all the revenue should be returned directly to U.S. citizens through equal lump-sum rebates. The majority of American families, including the most vulnerable, will benefit financially by receiving more in "carbon dividends" than they pay in increased energy prices.

Victor Hugo, the French poet and novelist said, "Nothing else in the world … not all the armies … is so powerful as an idea whose time has come." The carbon fee and dividend policy has broad bipartisan support from government officials, industry sectors, and environmental organizations. It is indeed a long-overdue idea.

DECARBONIZE THE GRID AND ELECTRIFY TRANSPORTATION.

I started this chapter with a quote from John Doerr (engineer, venture capitalist and chair of Kleiner Perkins) in his insightful TED Talk conversation with Hal Harvey (climate expert and CEO of Energy Innovation) where they discuss how to radically shift to a carbon-free economy by 2050. In their view, it's a simple formula: decarbonize the grid and electrify everything using renewable energy resources such as solar and wind, coupled with battery storage. I would modify their timeline slightly and aggressively target near 100 percent grid decarbonization and transportation electrification by 2035.

To make this happen, governments and policymakers must get on board immediately. Seventy-five percent of carbon emissions come from the twenty largest emitting countries and five sectors of the economy: power, transportation, buildings, agriculture, and industrial activities. To green these sectors at speed and scale requires robust government policy and financial incentives. In 2021, researchers at the University of California, Berkeley's Energy Innovation and GridLab published the *2035 Report 2.0*, in which they evaluated current battery and infrastructure costs to decarbonize the grid and electrify transportation. Their findings reveal that the transition to 100 percent electric cars sales by 2030 and all-electric truck sales by 2035, combined with a 90 percent decarbonized electric grid by 2035, could deliver $2.7 trillion in consumer savings through 2050. That equates to an average savings of $1,000 per household every year, while creating two million jobs in 2035 and reducing economy-wide carbon emissions 45 percent by 2030.

The economic and climate benefits of a clean electric grid and electrified transportation are clear. A growing chorus of voices are actively evangelizing about this message. Saul Griffith, author of *Electrify*, has been a long-time champion of the electrify everything movement, laying out both a technical and economic pathway to reach this goal. Today we are witnessing the reality of this transition as electric utilities and auto manufacturers embrace renewable energy and electric vehicles, with energy storage the 'holy grail' tying it all together. This is not a fad; it's a societal shift with long-term traction.

The pendulum appears to be shifting toward meaningful government action on climate policy. On the heels of the pandemic-triggered recession, in 2020 the European Union committed more than one trillion euros to fighting the climate crisis. Similarly, in 2021, a bipartisan US Congress passed and the US President signed into law a landmark $1.2 trillion climate-infrastructure plan. It will deliver $550 billion of new federal investments in America's infrastructure over five years, touching everything from bridges and roads to the nation's broadband, water, electric vehicles, and power transmission. Finally, in 2022 the US enacted the Inflation Reduction Act which authorizes $369 billion in climate and clean energy spending, making it the biggest and most consequential climate change bill ever passed by Congress.

All of this makes us hopeful that this time will be different. That this time, government leaders and industry will follow through on their pledges to decarbonize the global economy.

Reflections: Accountability

This chapter has clarified the meanings of government and corporate net-zero pledges, how to understand the metrics and data supporting the pledges, and how to keep both groups transparent and accountable. Ideally, carbon pledges should be set for 2035 or earlier and serve as a stepping-stone to net-zero. To prevent a worst-case climate scenario, there are three policy imperatives that all sovereign governments must enact and all corporate sectors must embrace: (1) adopt a carbon fee and dividend, (2) decarbonize the grid, and (3) electrify transportation.

Remember, each of you has a voice that deserves to be heard on the climate crisis. Whether that means emailing your government representatives, direct messaging a CEO on social media, or joining an organization that amplifies your voice and reflects your carbon values, your voice matters.

ACCOUNTABILITY CREED

I believe governments and companies are ultimately responsible for the impact of greenhouse gas emissions on human health and society.

Benchmark

The climate problem cannot be solved by individual consumption choices alone. Governments and corporations created this problem, and both must be held accountable for a solution. There are three policy imperatives that government must enact and corporations must embrace:

- Adopt a carbon fee and dividend.

- Decarbonize the electric grid.

- Electrify transportation.

Which policy do you feel would be most impactful today, and why? Resolve to take the following actions: (1) read Mission Economy by Mariana Mazzucato in the next twelve months; (2) track and understand the carbon emissions associated with ten products or services you use; and (3) contact a government official, direct message a CEO on social media, and join an organization that reflects your carbon values in the next three months.

CHAPTER 8

EQUITY

"It always seems impossible until it is done."
—NELSON MANDELA

"Star Trek was an attempt to say that humanity will reach maturity and wisdom on the day that it begins not just to tolerate but take a special delight in differences in ideas and differences in life forms."
—GENE RODDENBERRY

Nichelle Nichols had planned to leave *Star Trek* in 1967 at the end of its first season, wanting to return to music and theater. Prior to television, she had enjoyed a successful career touring the United States, Canada, and Europe as a singer with the Duke Ellington and Lionel Hampton bands.

That plan changed, however, when she learned that Martin Luther King Jr. was a *Star Trek* fan. King explained to Nichols, "You are our image of where we're going, you're 300 years from now, and that means that's where we are and it takes place now. Keep doing what you're doing, you are our inspiration."

As Nichols recounted, "*Star Trek* was one of the only shows that [King] and his wife Coretta would allow their little children to watch. And I thanked him and I told him I was leaving the show. All the smile came off his face. And he said, 'You cannot. Don't you understand for the first time we're seen as we should be seen. You don't have a Black role. You have an equal role.'"

Reach for the Stars

Few people know the story of Nichelle Nichols, an African American actor who landed the role of Lieutenant Uhura on the popular *Star Trek* franchise, and how that role opened our minds to a more equitable future. Star Trek showed us a future where equity was part of the solution to successful interstellar missions, just as today equity is key to successfully addressing climate change on Earth. Indeed, we can use lessons from *Star Trek* to shape our views on climate equity, in the spirit of Dr. King's words, "You are our image of where we're going, you're 300 years from now ... you are our inspiration."

Nichols played the role of Nyota Uhura, a translator and communications officer. As discussed under the kinship creed, effective communication is key to addressing our climate dilemma. The same is true when we talk about climate equity—it may require translating information for different perspectives on how carbon emissions impact our daily lives and our responsibility to reduce them. Different people and different communities will experience climate change and its effects in different ways. Europeans staring down fires and flooding do not face the same suffering as Pacific Islanders whose lives are destroyed by sea level rise and cyclones. Applying the Silver Rule from the ethics creed can help us see climate equity from the perspective of others.

The meaning of Nichols's character also adds weight to the symbolism of her role. Her given name *Nyota* is the Swahili word for "star." Her surname *Uhura*, is derived from the Swahili word *uhuru*, which means "freedom." Nichols explains in her memoir, *Beyond Uhura*, that the name was inspired by Robert Ruark's book *Uhuru*, which she had with her on the day she read for the part. Clearly, *Star Trek* creator Gene Roddenberry was both inspired and intentional when he created this character. It's not coincidental that escaped black slaves followed the North Star to freedom. With a mere name choice, Roddenberry is showing us what equity looks like. Subconsciously, he's freeing our minds to see equity.

In a similar way, we need to *see* climate equity to understand it. In this book, we believe in economic and social climate equity for all humans. We see climate equity as transcending race, wealth, nationality, gender, and political barriers. Equity does not allow the people least responsible for carbon emissions to bear the brunt of the climate crisis. Rather, we support fair and equitable access to climate investments and economic opportunities for all demographic groups. This is the equity creed.

As you can tell from my opening homage to *Star Trek* and earlier references to *Star Wars*, I am a big fan of science fiction. As a genre, science fiction can help people see the future in ways that defy the present. That's what makes Gene Roddenberry's work timeless. *Star Trek* allows us to see what humanity can become—an advanced society of humans where equity is the rule. Nyota Uhura is our future.

Many of the conflicts and political dimensions presented in *Star Trek* were simply allegories of contemporary cultural realities. The original *Star Trek* series addressed sociopolitical issues of the 1960s, just as later iterations and spin-offs have confronted issues of their times. Issues depicted in the various series include war and peace, authoritarianism, imperialism, economics, racism, religion, human rights, sexism, and the role of technology. I have no doubt that if Roddenberry had started his franchise today, human impacts on climate would have been a central theme in his universe.

Roddenberry created *Star Trek* to show a future of what human society might become if people would learn from the lessons of the past. He purposely made Lieutenant Uhura a central figure to show us what equity should look like. For kids who looked like me, seeing her helped me see myself in that future universe. Sulu, Chakotay, Janeway, Bashir, Burnham and other characters did the same for kids who looked different from me.

Today, when I attend climate conferences and negotiations where scientists and policy experts are shaping the world's climate equity future, I see few people who look like me. The people who represent us at the table matter as much as the policy decisions they render.

To further stress the value of representation, let's revisit our discussion about *Star Wars* and consider a broader understanding of the term Jedi. Coded in the title is the acronym justice, equity, diversity, and inclusion. Whether done purposely or not, the Jedi Order manifests all four qualities:

- justice (the Jedi Council seeks solutions that are right and fair)

- equity (Jedi seek balance in the Force)

- diversity (Jedi are multispecies, force-sensitive beings)

- inclusion (Jedi are men, women, and children)

Of these four qualities, I believe equity is central. In fact, the core Jedi prophecy foretold "one who would bring balance to the Force." As discussed in the ethics creed, the Force is a source of universal power where balance represents peace. In a similar way, humans must achieve equity to bring balance to the climate. When humans learn to see one another as equals, we will move away from the zero-sum mindset that has plagued our species and now threatens the planet. We will do whatever it takes to protect our shared home, our blue origin.

This is why I believe that equity is central to mitigating climate change. In no way does this viewpoint diminish justice, diversity, or inclusion—all are important. However, in this book, equity is the first priority. Ultimately, whether you're a Padawan just starting this journey or an experienced Knight, your goal is to become a Jedi Master. It is not something that just happens. It has to be lived. It has to be intentional. There is no try.

Now, let's use these lessons from science fiction to examine the real challenges associated with climate equity. Across the planet, there exists an ideological divide over the climate equity problem. People living in the Northern Hemisphere tend to view climate equity from an accounting perspective—set a numeric goal and reduce carbon intensity or greenhouse gas emissions over time. In the Southern Hemisphere, equity is centered on who is causing the climate crisis and who will bear the burden.

In his seminal paper "Equity in Climate Change: The Great Divide," Benito Müller finds that one of the root causes of this divide is a fundamental difference in the perception of climate change itself. "In the industrialized nations of the North, there is a widely held 'ecological view' of the problem. Its essence is that of wrongful actions against 'Nature.' The chief victim from this perspective is Nature, while humanity's role is primarily that of culprit. Environmental integrity ('to do justice to Nature') is the overriding moral purpose. The reality in the South is quite different: climate change has primarily come to be seen as a human welfare problem. The harm is against humans, it is largely other-inflicted, and it is not life-*style*-, but *life*-threatening. The chief victim of climate change is not 'Nature,' but people, and the paramount inequity is one between human victims and human culprits." Müller first published his observations in 2002, and they remain true today.

This ideological divide is further complicated by an economic divide. The overwhelming majority of wealthy humans live in the Northern Hemisphere, while the poorest humans are almost entirely in the countries of the Southern Hemisphere. The "global middle class" are in fact very poor—half of them earn less than twenty dollars a day and many earn less than three dollars a day. They are almost all in the Global South. Broadly speaking, industrial countries of the Global North, with a combined 20 percent of global population, are responsible for approximately 80 percent of all industrial greenhouse gas emissions. These countries have grown immensely wealthy and powerful, in part because of the systematic plunder of the Global South, its people, and their resources.

Unfortunately, we see these same patterns *within* industrialized nations. In August 2005, Hurricane Katrina, a Category 5 storm at its peak intensity, caused an estimated $160 billion in damage and displaced more than a million residents of the US Gulf Coast. The unusually large hurricane was also among the country's deadliest, killing an estimated 1,833 people across seven states, from Florida to Ohio. But it was the devastation experienced in New Orleans that shocked the nation and continues to haunt the city. Journalists documented which residents took the brunt of the disaster and whose lives were the most profoundly upended and lost. Not surprisingly, they were primarily low income, older, and people of color. This follows the common pattern of the poor and marginalized communities being forced to live in the less desirable parts of cities, lacking the necessary infrastructure and community resources to promote resiliency and protect against severe weather and climate impacts.

In another incident, the town of Paradise, California, burned to the ground in 2018, killing eighty-five people. The wildfire, known as the Camp Fire, was the deadliest and most destructive in California history, and the most expensive natural disaster on the planet in 2018 in terms of insurance losses. Before the fire, Paradise was a poor, racially diverse community with a median annual income of less than $50,000—below the national average. In Paradise, climate change combined with economic disadvantage to create a deadly situation. This is not to say that only the poor were impacted by the Camp Fire; we know that the wealthy were affected too. However, marginalized people historically have fewer resources to bounce back, leading to a higher risk of financial ruin.

Sadly, these kinds of scenarios are increasingly common as the climate crisis accelerates. Scientists estimate that climate change has increased wildfire risk by 500 percent, compared to risk levels in the twentieth century. Climate change also is making heat waves hotter, hurricanes stronger, and droughts longer. And the pace of warming continues to increase—humans have already warmed the planet by 1 degree Celsius, which is halfway to the danger level that world leaders vowed to avoid under the Paris Agreement.

So, yes, we need to address the social consequences of climate equity. Because we all have the right to clean water and clean air. Access to these resources is threatened by the climate crisis. If we don't take an equitable approach to climate, we are essentially telling billions of people that their lives don't matter as much as our lifestyles.

The Economics of Equity

At the beginning of this book, I shared my view that equity in the context of carbon and climate has two sides—social and economic. I believe we need both social and economic equity to achieve a fair, sustainable future for all humans. We need broader economic opportunity, especially for marginalized people. So let's talk about sharing the wealth.

In the United States, most wealth creation is associated with investing in financial markets. There is a movement underway in the financial markets that has the potential to advance the importance of climate and economic equity. It is called environmental, social, and governance (ESG) investing. ESG is a form of sustainable investing in which investors use corporate pledges, discussed under the accountability creed, as a metric to choose where to invest their money. A report by McKinsey Sustainability defines the ESG metrics as follows:

E, environmental criteria

Includes the energy a company takes in and the waste it discharges, the resources it needs, and the consequences for living beings as a result. Not least, *E* encompasses carbon emissions and climate change. Every company uses energy and resources; every company affects, and is affected by, the environment.

S, social criteria

Addresses the relationships a company has and the reputation it fosters with people and institutions in the communities where you do business. S includes labor relations and diversity and inclusion. Every company operates within a broader, diverse society.

G, governance

Is the internal system of practices, controls, and procedures a company adopts in order to govern itself, make effective decisions, comply with the law, and meet the needs of external stakeholders. Every company, which is itself a legal creation, requires governance.

According to a 2021 Bloomberg Intelligence report, global ESG assets are on track to exceed $53 trillion by 2025, representing more than a third of the projected $140.5 trillion in total assets under management. Aside from the benefits of creating a more ethical portfolio, there is evidence that ESG investments deliver similar returns as traditional investments and potentially carry less risk.

Unfortunately, many companies don't make the grade when it comes to meeting the lofty goals of ESG investments. The first area of concern is the scarce number of directors with environmental credentials.

According to a study released by New York University's Stern Center for Sustainable Business, as recently as 2018, of the 1,188 directors at one hundred of the biggest US companies, just 6 percent had "relevant credentials" on environmental matters and only 0.3 percent had expertise in either climate (0.2 percent) or water (0.1 percent) issues. A second area of concern is the use of ESG as a marketing gimmick without any substance or investment.

This is especially alarming given that about 1,541 companies have announced net-zero targets covering 3.5 billion metric tons of greenhouse gas emissions, according to a report by Data-Driven EnviroLab and NewClimate Institute, and just a small fraction of them have concrete plans in place that will get them to their targets by 2050. Finally, we cannot forget the meager number of minorities and women appointed to corporate boards. You get the picture. Do your homework.

If you choose to invest in the financial markets, the Carbon Creed compels you to align your investments with your ethics. Investing is a lifetime practice, not a singular action. The decisions you make now based on your beliefs and values will determine your returns for years to come. Thus, it is important that you have a plan to identify sustainable assets. I recommend that you consider joining an ESG focused investment club or subscribing to a newsletter. Take your time to find a good community and content fit. When it comes to choosing assets (human and financial), diversification is the key to reducing risk and enhancing returns. The wisdom of crowds can be a powerful financial tool, especially when the crowd includes others who embrace the Carbon Creed.

While ESG investing is an option for some people, it is out of reach for many. The World Bank says, "The number of poor worldwide remains unacceptably high, and it is increasingly clear that the benefits of economic growth have been shared unevenly across regions and countries." We can trace much of the global wealth gap to the institutions of slavery, colonialism, and imperialism, which sought to concentrate the world's wealth in the hands a few privileged humans.

Today, we have the chance to change the arc of history to allow humans of all races, nationalities, genders, and creeds to share in the opportunities of economic wealth creation. The blockchain and virtual currency movement is poised to reset the way we define the social construct of money.

Satoshi Nakamoto, the developer and creator of Bitcoin, explained in his seminal 2008 white paper that crypto currencies could provide a significant benefit to society by overcoming lack of social trust and increasing the access to financial services. This could be a game changer for marginalized people and nations that have historically lacked sufficient access to capital and credit facilities. Even people without bank accounts can now use cryptocurrencies if they have digital wallets. Cheaper transaction fees can enable people in developing nations to own property with fewer financial burdens. For some migrants, cryptocurrencies can serve as a remittance tool, allowing them to transfer money between countries with no or very low costs of transfer and allowing for exchanges between currencies with no markup.

Let me be clear. This is not an endorsement to invest in crypto assets. What I am saying is if you care about economic equity and the future, you need to get educated about how this new asset class could potentially redefine and democratize global wealth.

As followers of the Carbon Creed, you should also be aware of the energy and climate implications associated with cryptocurrencies. On the surface, it may seem unlikely that digital currencies would have a significant carbon footprint. But cryptocurrency mining, the process that adds new units of crypto into circulation, can be highly carbon intensive, unless the electricity is generated using clean energy sources like solar and wind. Do your homework. Align your values with your investments by understanding the difference between *proof of work* (energy intensive) and *proof of stake* (energy efficient) currencies.

Furthermore, when it comes to decarbonizing global commerce, digital currencies and blockchain could provide the missing link. Kim Stanley Robinson's recent science fiction novel *Ministry for the Future* introduces a new variable into the carbon equation: a reward for doing good. In his novel, Robinson's public servant hero infuses muscle into the Paris Agreement by persuading officials at central banks to support a new currency called the carbon coin.

The carbon coin becomes a pivotal incentive for moving governments, corporations, and individuals down the path of decarbonization. Each digital coin is produced when a ton of carbon is eliminated from the atmosphere—such as kelp forests capturing carbon from the air and sea, petroleum companies leaving reserves in the ground, or subsistence farmers rewilding their fields.

Although it may sound like pure science fiction, the carbon coin concept is drawn from real projects aimed at creating a reward for reducing the risks of climate change. The Global Carbon Reward and the Energy Web Token are both viable contenders for making the carbon coin a reality. To be clear, the proposed carbon coin is a digital currency, not a cryptocurrency. It is important to understand the differences between the two. Both can be blockchain based. However, digital currencies are centralized, meaning that transaction within the network is regulated in a central location, like a bank. Cryptocurrencies are mostly decentralized, and the regulations inside the network are governed by the community.

While it is impossible to predict how crypto currencies, digital currencies, blockchain, and climate will ultimately converge in the future, we still need to price decarbonization into our economic system. I believe that we will eventually put a price on carbon, and that price will likely be denominated using a digital or cryptocurrency. That much seems certain.

Reflections: Equity

This chapter has examined the social and economic dimensions of climate equity. Climate equity transcends race, wealth, nationality, gender, and political barriers. It does not allow the people least responsible for carbon emissions to bear the brunt of the climate crisis. Rather, it supports fair and equitable access to climate investments and economic opportunities for all demographic groups.

It is my hope that you will use this insight and understanding to boldly create a new lens for climate equity. Educate yourself about ESG investing and the shift toward sustainable assets. Investigate the risks and opportunities associated with blockchain, cryptocurrencies, and digital currencies as tools to price decarbonization and create wealth.

EQUITY CREED

I believe in climate equity and opportunity
for all humans.

Benchmark

Climate equity is an opportunity to reset humanity's social and economic divisions by redefining the relationship between wealth and the planet.

Which dimension of climate equity are you more drawn to—social or economic? Be resolved to take these actions: (1) read Ministry of the Future by Kim Stanley Robinson and The Nutmeg's Curse by Amitav Ghosh in the next twelve months, (2) view sentient life as precious, and (3) align your carbon values with your investments.

CONCLUSION

"You are not Atlas carrying the world on your shoulder. It is good to remember that the planet is carrying you."
—VANDANA SHIVA

You've reached the final chapter of this book, and your mind is probably racing with ideas. We've covered a lot of ground and you're likely wondering, "With so much information, where do I begin?"

A good place to start is a review of the creeds presented in this book. It will help you assess where you are on the path to decarbonization. Let's refresh our minds with the following chapter summaries.

BELIEFS.

You understand the seven core beliefs that make up the Carbon Creed. The Carbon Creed is your starting point on a journey. It is a *pathway* to the beliefs, values, and practices that will inspire you and others to decarbonize. Through the repetition of positive affirmations, you can train your subconscious mind to become carbon conscious.

ETHICS.

You can see the wisdom of ethical rules and will apply the Silver Rule (do no harm) and Golden Rule (do good) as your compass to decarbonization. Together they embody the moral imperative of the entire creed. If you start by doing no harm to others, and do good when you have the ability, then your carbon virtues, values, and actions will be aligned. Embrace the Hippocratic Oath for climate.

MINDSET.

Your careful study of the Wizard and the Prophet analogy has revealed your mindset and leanings toward carbon innovation (Wizard) or conservation (Prophet). You recognize that these are complimentary mindsets. *Innovation* supports the benevolent use of data and technology to decarbonize society, while *conservation* supports minimalism and a shared multigenerational responsibility to steward the planet. You don't have to be wealthy or famous to be a carbon influencer.

SCIENCE.

You believe the scientific consensus that the earth's climate is warming and human activities are the primary cause. You understand that we are living in the Anthropocene, a time when humans are the dominant force on Earth, terraforming and changing the planet's ecosystems and processes. There are both ecological and psychological dimensions to the science of climate change, and both are important. Ultimately, you should feel hopeful because you know that humanity possesses the technology to avoid climate disaster—if we act now.

KINSHIP.

You understand that translating climate change from a scientific reality to a societal imperative is extremely difficult. The failure to communicate climate change effectively has divided humans into tribal factions. To find common ground on carbon and climate, you must learn to listen. Remember, facts don't change minds, people do. Welcome people with different viewpoints and backgrounds into your fold. Be kind and make them kin.

HABITS.

You believe that our lives are not shaped by big decisions or events, but by our daily habits. You understand that rituals can transform your life by turning everyday habits and routines into meaningful practices that decarbonize your life and engage others. You have discovered the five carbon-conscious rituals: sacred text, sabbath rest, sanctuary, scribe, and supper. Identify the rituals and practices that inspire, make you more carbon conscious, and motivate you to action.

ACCOUNTABILITY.

You believe we are witnessing a global shift by governments and corporations to adopt net-zero carbon pledges, and it is up to individuals like you to keep them transparent and accountable. You understand that carbon pledges should require actual carbon emissions reductions, not just offsets. Policy priorities must include a price on carbon, transportation electrification, and a decarbonized grid. Government policy and corporate behavior created the climate crisis, and both are necessary to solve it.

EQUITY.

You have learned that equity in the context of carbon and climate has both social and economic dimensions.

It should support fair and equitable access to climate investments and economic opportunities for all demographic groups. It should also ensure the people least responsible for carbon emissions are not the most impacted by climate change. Equity is the final frontier.

How does it feel to absorb all this information? Has it effected your thinking about carbon and climate? Are you resolved to act on what you've learned? I'm sure you have lots of mixed feelings as you contemplate these questions, and that's to be expected. This book appeals to people for different reasons, but key among them is the focus on candid advice about shaping your beliefs, values, and practices to become carbon conscious. I firmly believe that if you really want to defeat catastrophic climate change, you must decarbonize your life first. We know from the accountability creed that you are not responsible for the climate crisis, and ultimately governments and corporations will have to fix this problem at scale. But that doesn't mean you should feel helpless. Taking small steps to lower your carbon footprint can ease climate anxiety by giving you back a sense of power and control. Even the smallest actions matter. But how do you start? We started this in the habits creed, I'm going to show you more now.

Carbon Conscious Mornings

I created **Carbon Conscious Mornings**[SM], a plan that will put you on the path to a low-carbon lifestyle instantly. "But why mornings?" you may ask. For most of my life, I never considered myself a morning person. In fact, I hated getting up early. What changed? For me it was discovering that many successful people start their days with early morning rituals, so I decided to do likewise but make my morning rituals carbon conscious. Today, I easily wake up around 6:00 a.m. every morning, excited to begin a new day. I live by the mantra: *A morning ritual sets the tone for the whole day, and if you do each day right, you'll do life right.*

Let me share with you five morning rituals to help reduce your carbon footprint from the moment you get out of bed:

1. SHOWER RITUAL

There's nothing quite as soothing as a hot shower to start the day, especially when the weather gets colder. However, longer showers (e.g., 20 minutes) use more energy, which leads to a higher carbon footprint. Why not make your shower ritual carbon conscious by limiting yourself to ten minutes or less? To enhance the ritual, create a shower playlist as a timer and challenge yourself to be out before it's finished. I also recommend switching to a low-flow shower head to reduce your carbon footprint. A standard shower head uses around eighteen liters per minute, while a low-flow unit only uses nine liters per minute. They work in the exact same way as a regular shower head and provide the exact same shower experience. I installed a low-flow shower head at home, and it elevated my routine shower to a spa-like experience.

2. COFFEE RITUAL

I'm not a coffee snob, but I cherish my morning coffee ritual. The whole process of making and drinking that first cup is a pure delight to my senses. Coffee, which is a plant-based beverage, is also a great carbon-conscious choice to start your day. Every cup of coffee will leave you with grounds, which often go to a landfill. Why not save those grounds and use them in your garden? That's something we have done. Another carbon-conscious choice would be to go dairy free. Drinking black coffee is the healthiest way to go, but no one would object if you used a plant-based coffee creamer. That's what I do. The variety of nondairy creamers is long and delicious: almond, oat, soy, cashew, and coconut to name a few. Each varies in terms of taste, mouthfeel, and nutritional properties, so find one that fits your preference.

3. BREAKFAST RITUAL

Most scientists agree that the fewer animal products we consume, the more sustainable our diet and the healthier we will be. Studies show that beef, lamb, and animal products like dairy generate the largest amount of food-related greenhouse gas emissions. While you may not be ready to commit to a 100 percent vegetarian diet, cutting back on animal products can lower your carbon footprint significantly. For example, one popular way to reduce meat is to remove it from your breakfast. You can do this every morning or on set days.

There are many excellent choices when it comes to plant-based breakfast. Whole grain cereals, oatmeal, waffles, pancakes, tofu scramble, or a veggie burrito to name a few. I have adopted a plant-based breakfast ritual for weekdays. My go-to meal is oatmeal with cinnamon, topped with nuts and fresh berries. You can change the types of nuts and berries daily to add some variety. My second favorite option is pumpkin seed and flax granola topped with blueberries. We try to always keep at least a thirty-day stash in the cupboard.

4. READING RITUAL

After finishing my shower and coffee rituals, I dive right into my morning ritual of reading the daily newspaper. I start my day browsing *The New York Times*, *The*

Wall Street Journal, *The Washington Post*, and the local paper all on my iPad. As I suggested in the habit creed chapter, you can make this choice carbon conscious by changing your subscriptions from paper to electronic, which I did more than seven years ago. I also read with an eye to carbon and climate stories, and I share them with others on social media along with my personal insights. I find this ritual mentally stimulating and much healthier than bingeing morning cable news.

5. FITNESS RITUAL

I am a strong believer in fitness and exercise, especially in the outdoors. Gyms are great and I belong to one but there are more carbon-conscious benefits and options when you are outdoors. For example, walking or hiking in a forest can bolster your immune system, ward off viral infections and disease, and drive down blood pressure and anxiety. An outdoor fitness ritual could also allow you to convert a vehicle commute into a bike ride or walk. These are just a few ways to stay active, healthy, and fit that are carbon conscious and don't require a gym or treadmill.

Of the five morning recommendations, the fitness ritual may be the hardest to stick with. While not required, I suggest investing in a wearable fitness tracker. I purchased an Apple Watch solely for this reason and it has changed my life. It allows me to set a daily fitness goal, monitors my activities and provides valuable health data (e.g., calories, heart rate, respiratory rate, blood oxygen, and step counts). It also prompts me to practice mindfulness exercises throughout the day.

***Carbon Conscious Mornings*^{SM}** is a great tool to jumpstart your low-carbon lifestyle. It is meant to serve as a turnkey solution, so you don't have to create a custom plan on your own. However, if you have other ritual preferences, feel free to use them—there are many different ideas presented throughout this book. Just be sure to focus on one ritual at a time, keep them small to start and set a preferred time of day to practice.

Keep Helping Others

In these final words, if there is anything I can do to support you or add value to your decarbonization journey in any way, please let me know. You don't have to carry the weight of the world on your shoulders. I am here to help.

I'm always grateful to connect with other carbon-conscious people and find it truly awesome to hear from people who have read my book, subscribe to my newsletter, or engage with me on social media. We are building a community based on the ideas and teachings found in this book. Come join us at www.carboncreed. com. I look forward to hearing from you, learning new ways to add value to your life, and helping you reach your goals.

Now, I have one last thing to ask of you. If this book has added value to your life, if you feel like it's made you a better communicator of the climate problem and how to solve it, and you see the Carbon Creed as a platform for you to take your sustainability goals in life to the next level, I'm hoping you'll do something special for someone you care about: Share this book with them. Let them borrow your copy. Ask them to read it. Or better yet, give them their own copy as a gift.

You don't have to wait for a special occasion to share the Carbon Creed with others; it's fine just to say, "I'm giving you this because I care about you, and I want to help you live the most sustainable life possible. I hope you enjoy reading it."

Please spread the word. Thank you so much.

References

(Sources are arranged alphabetically by chapter.)

INTRODUCTION

Heglar, Mary Annaïse. "We Can't Tackle Climate Change Without You." *Wired*. April 1, 2020. https://www.wired.com/story/what-you-can-do-solve-climate-change.

Hower, Mike. "Why We Need to Rethink These Three Climate Metaphors." *GreenBiz*. January 11, 2019. https://www.greenbiz.com/article/why-we-need-rethink-these-three-climate-metaphors.

Litvinova, Daria, and Seth Borenstein. "The Arctic Is on Fire: Siberian Heat Wave Alarms Scientists." *AP News*. June 23, 2020. https://apnews.com/article/health-fires-ap-top-news-international-news-moscow-39c5b62946a0951cf0d98e4f952f47fd.

Pappas, Stephanie. "'Zombie' Anthrax Outbreak in Siberia: How Does It Kill?" *Live Science.* August 2, 2016. https://www.livescience.com/55621-zombie-anthrax-kills-in-siberia.html.

Sakai, Ko. "Siberia's Warming Shows Climate Change Has No Winners." *Nikkei Asia.* March 14, 2021. https://asia.nikkei.com/Spotlight/Comment/Siberia-s-warming-shows-climate-change-has-no-winners.

University of Colorado at Boulder. "The Arctic Is Burning in a Whole New Way: Widespread Wildfires in the Far North Aren't Just Bigger; They're Different." *ScienceDaily.* September 28, 2020. www.sciencedaily.com/releases/2020/09/200928155746.htm.

CHAPTER 1: BELIEFS

Davis, Jarrod. "Values vs. Beliefs." Barrett Values Centre (blog). https://www.valuescentre.com/values-vs-beliefs.

"How to Reprogram Your Subconscious Mind for Success and Happiness." *Great Performers.* Accessed September 12, 2022. https://greatperformersacademy.com/habits/how-to-reprogram-your-subconscious-mind-forsuccess-and-happiness.

Merriam-Webster Online, s.v. "credo." https://www.merriam-webster.com/dictionary/credo.

"Nicene Creed." *Wikipedia*. Accessed August 31, 2022. https://en.wikipedia.org/wiki/Nicene_Creed.

"The True Story of The American's Creed." *This Family Blog*. https://thisfamilyblog.com/the-story-of-the-americans-creed.

CHAPTER 2: ETHICS

Humphreys, Joe. "Unthinkable: Which 'Golden Rule' of Ethics Is Best, The Christian or Confucian?" *The Irish Times*. January 31, 2014. https://www.irishtimes.com/culture/unthinkable-which-golden-rule-of-ethics-is-best-the-christian-or-confucian-1.1674003.

"The Force." *Wikipedia*. Accessed August 31, 2022. https://en.wikipedia.org/wiki/The_Force.

"The Golden Rule." *Wikipedia*. Accessed February 28, 2020. https://en.wikipedia.org/wiki/Golden_Rule.

Gonen, Jordan. "The Golden Rule vs. The Silver Rule." *Gonen Blog*. October 30, 2018. https://gonen.blog/the-golden-rule-vs-the-silver-rule.

Esparza, Daniel. "Is It 'Do Unto Others' or 'Don't Do Unto Others'? What The Golden Rule Demands." *Aleteia*. February 24, 2019. https://aleteia.org/2019/02/24/is-it-do-unto-others-or-dont-do-unto-others-what-the-golden-rule-demands.

"The Analects of Confucius." *Confucious-1.com*. *http://www.confucius-1.com/analects*.

Miller, James. "How Confucianism Could Curb Global Warming." *The Christian Science Monitor*. June 26, 2009. https://www.csmonitor.com/Commentary/Opinion/2009/0626/p09s01-coop.html.

"Hippocratic Oath." *Wikipedia*. Accessed August 31, 2022. https://en.wikipedia.org/wiki/Hippocratic_Oath.

Pearce, Fred. "After the Coronavirus, Two Sharply Divergent Paths on Climate." *Yale Environment 360*. April 7, 2020. https://e360.yale.edu/features/after-the-coronavirus-two-sharply-divergent-paths-on-climate.

CHAPTER 3: MINDSET

Johnson, Nathanael. "Wizards and Prophets Face Off to Save the Planet." *Grist*. February 9, 2018.

https://grist.org/article/wizards-and-prophets-face-off-to-save-the-planet.

Mann, Charles C. "Can Planet Earth Feed 10 Billion People?" *The Atlantic*. March 2018. https://www.theatlantic.com/magazine/archive/2018/03/charles-mann-can-planet-Earth-feed-10-billion-people/550928.

Mann, Charles C. *The Wizard and the Prophet: Two Remarkable Scientists and Their Dueling Visions to Shape Tomorrow's World*. New York: Alfred A. Knopf, 2018.

Rome, Adam. "The Bitter Brawl Over Humanity's Future." *Nature* 553 (January 10, 2018): 152-153. Accessed August 31, 2022. https://www.nature.com/articles/d41586-018-00111-8. doi: *10.1038/d41586-018-00111-8*.

Sylvia, Tim, and John Fitzgerald Weaver. "A Way to Achieve Greater Than 100% Solar Power in the U.S., Without Sacrificing Arizona." *pv magazine USA*. December 12, 2019.

CHAPTER 4: SCIENCE

"Alexander von Humboldt." *Wikipedia*. Accessed August 31, 2022. https://en.wikipedia.org/wiki/Alexander_von_Humboldt.

"Svante Arrhenius." *Wikipedia*. Accessed August 31, 2022. https://en.wikipedia.org/wiki/Svante_Arrhenius.

de Pencier, Nicholas, Edward Burtynsky, and Jennifer Baichwal. "The Anthropocene Project." https://theanthropocene.org.

Ellis, Erle. "The Long Anthropocene." *The Breakthrough*. May 1, 2013. https://thebreakthrough.org/articles/the-long-anthropocene.

Grabianowski, Ed. "Terraforming Earth: How to Wreck a Planet in 3,000 Years (Part 1)." *Wired*. September 24, 2010. https://www.wired.com/2010/09/terraforming-part-1.

Intergovernmental Panel on Climate Change. *Climate Change 2022: Impacts, Adaptation and Vulnerability*. February 28, 2022. https://www.ipcc.ch/report/ar6/wg2.

Marks, Elizabeth, Caroline Hickman, Panu Pihkala, Susan Clayton, Eric R. Lewandowski, Elouise E. Mayall, Britt Wray, Catriona Mellor, and Lise van Susteren. "Young People's Voices on Climate Anxiety, Government Betrayal and Moral Injury:

A Global Phenomenon." September 7, 2021. https://ssrn.com/abstract=3918955.

Neukom, Rachel, Nathan Steiger, Juan José Gómez-Navarro, Jianghao Wang, and Johannes P. Werner. "No Evidence for Globally Coherent Warm and Cold Periods Over the Pre-Industrial Common Era." *Nature* 571 (July 24, 2019): 550–554. Accessed August 31, 2022. doi: 10.1038/s41586-019-1401-2.

PAGES 2k Consortium. "Consistent Multidecadal Variability in Global Temperature Reconstructions and Simulations Over the Common Era." *Nature Geoscience* 12 (July 24, 2019): 643–649. Accessed August 31, 2022. https://www.nature.com/articles/s41561-019-0400-0. doi: 10.1038/s41561-019-0400-0.

Stoknes, Per Epsen. "How to Transform Apocalypse Fatigue into Action on Global Warming." Filmed November 17, 2017, in New York City. TED video, 14:51. https://www.ted.com/talks/per_espen_stoknes_how_to_transform_apocalypse_fatigue_into_action_on_global_warming?language=en.

Y Combinator. "Carbon Removal Technologies." http://carbon.ycombinator.com.

Yuhas, Daisy. "A Year of the Pandemic: How Have Birds and other Wildlife Responded?" *Audubon Magazine*. March 9, 2021. https://www.audubon.org/news/a-year-pandemic-how-have-birds-and-other-wildlife-responded.

CHAPTER 5: KINSHIP

Clear, James. "Why Facts Don't Change Our Minds." *JamesClear.com*. September 10, 2018. https://jamesclear.com/why-facts-dont-change-minds.

Hendricks, Rose. "Communicating Climate Change: Focus on the Framing, Not Just the Facts." *The Conversation*. March 5, 2017. https://theconversation.com/communicating-climate-change-focus-on-the-framing-not-just-the-facts-73028.

Leiserowitz, Anthony, Edward Maibach, Seth Rosenthal, John Kotcher, Parrish Bergquist, Matthew Ballew, Matthew Goldberg, and Abel Gustafson. *Climate Change In the American Mind: November 2019*. Yale University and George Mason University. New Haven: Yale Program on Climate Change Communication, 2019.

Markowitz, Ezra, and Adam Corner. "Climate Change Is Really About Prosperity, Peace, Public Health and Posterity—Not Saving the Environment."

The Conversation. September 27, 2019. https://theconversation.com/climate-change-is-really-about-prosperity-peace-public-health-and-posterity-not-saving-the-environment-120476.

Matusek, Sarah. "Why You Should Talk About Climate Change—Even If You Disagree." *The Christian Science Monitor*. December 5, 2019. https://www.csmonitor.com/Environment/2019/1205/Why-you-should-talk-about-climate-change-even-if-you-disagree.

Webster, Robin. "How to Talk About Climate Change with Family and Friends Over the Holidays." *Climate Home News*. December 12, 2019. https://www.climatechangenews.com/2019/12/20/talk-climate-change-family-friends-holidays.

CHAPTER 6: HABITS

Compton, Julie. "How our Diets Impact Climate Change—And What We Can Do About It." *NBC News BETTER*. August 12, 2019. https://www.nbcnews.com/better/lifestyle/how-our-diets-impact-climate-change-what-we-can-do-ncna1041301.

"What's the Carbon Footprint of my Smartphone?" *Honest Mobile*. August 25, 2020. https://

honestmobile.co.uk/2020/08/25/whats-the-carbon-footprint-of-my-smartphone.

Kashdan-Brown, Jessica. "How Journaling Can Help Us Process the Climate Crisis." *The Climate Journal Project.* January 30, 2021. https://theclimatejournalproject.medium.com/6-benefits-of-journaling-671bcd3e0a06.

Kuile, Casper ter. *The Power of Ritual: Turning Everyday Activities into Soulful Practices.* New York: HarperOne, 2020.

Schaper, Donna. "When a Ritual Works: Reflections on The People's Climate March." *EarthBeat.* September 24, 2014. https://www.ncronline.org/blogs/earthbeat/eco-catholic/when-ritual-works-reflections-people-s-climate-march.

Razzetti, Gustavo. "The Power of Rituals: How to Build Meaningful Habits." *Medium.* February 11, 2019. https://medium.com/personal-growth/the-power-of-rituals-how-to-build-meaningful-habits-697afb0ba0da.

CHAPTER 7: ACCOUNTABILITY

"Economists' Statement on Carbon Dividends: Bipartisan Agreement on How to Combat Climate Change."

The Wall Street Journal. January 6, 2019. https://www.wsj.com/articles/economists-statement-on-carbon-dividends-11547682910.

Byskov, Morten Fibieger. "Focusing on How Individuals Can Stop Climate Change Is Very Convenient for Corporations." *Fast Company.* January 11, 2019. https://www.fastcompany.com/90290795/focusing-on-how-individuals-can-stop-climate-change-is-very-convenient-for-corporations.

Chemnick, Jean. "Russia's Climate Posture Offers Clues on Its Ukraine Mindset." *ClimateWire.* February 24, 2022. https://www.eenews.net/articles/russias-climate-posture-offers-clues-on-its-ukraine-mindset.

Derviş, Kemal, and Sebastian Strauss. "The Decarbonization Paradox." *Project Syndicate.* February 17, 2021. https://www.project-syndicate.org/commentary/rapid-decarbonization-may-be-both-impossible-and-inevitable-by-kemal-dervis-and-sebastian-strauss-2021-02.

Goldman School of Public Policy, University of California, Berkeley. *2035: The Report.* April 15, 2021. https://www.2035report.com.

Griffith, Saul. *Electrify: An Optimist's Playbook for Our Clean Energy Future.* Cambridge: MIT Press, 2022.

Leahy, Stephen. "Most Countries Aren't Hitting Paris Climate Goals, and Everyone Will Pay the Price." *National Geographic.* November 5, 2019. https://www.nationalgeographic.com/science/article/nations-miss-paris-targets-climate-driven-weather-events-cost-billions.

McDonnell, Tim. "How to Call BS on Corporate Climate Pledges." *Quartz.* September 29, 2020. https://qz.com/1909846/how-to-call-bs-on-corporate-climate-pledges.

Osusky, Dan. "How to Determine If Corporate Climate Commitments Are Legitimate." *Fast Company.* January 10, 2020. https://www.fastcompany.com/90449900/how-determine-if-corporate-climate-commitments-are-legitimate.

Patnaik, Sanjay, and Kelly Kennedy. "Why the US Should Establish a Carbon Price Either Through Reconciliation or Other Legislation." *Brookings.* October 7, 2021. https://www.brookings.edu/research/why-the-us-should-establish-a-carbon-price-either-through-reconciliation-or-other-legislation.

CHAPTER 8: EQUITY

Bloomberg Intelligence. "ESG Assets May Hit $53 Trillion by 2025, a Third of Global AUM." *Bloomberg Professional Services.* February 23, 2021. https://www.bloomberg.com/professional/blog/esg-assets-may-hit-53-trillion-by-2025-a-third-of-global-aum.

Data-Driven EnviroLab & NewClimate Institute. *Accelerating Net Zero: Exploring Cities, Regions, and Companies' Pledges to Decarbonise.* September 2020. http://datadrivenlab.org/wp-content/uploads/2020/09/Accelerating_Net_Zero_Report_Sept2020.pdf.

EmmyTVLegends.Org. "Nichelle Nichols on How Dr. MLK, Jr. Dissuaded Her from Quitting Star Trek." FoundationINTERVIEWS. January 8, 2013. Video interview excerpt, 12:07. https://youtu.be/pSq_UIuxba8.

Henisz, Witold, Tim Koller, and Robin Nuttall. "Five Ways that ESG Creates Value." *McKinsey Quarterly.* November 2019. https://www.mckinsey.com/~/media/McKinsey/Business%20Functions/Strategy%20and%20Corporate%20Finance/Our%20Insights/Five%20ways%20that%20ESG%20creates%20value/Five-ways-that-ESG-creates-value.ashx.

Holtmeier, Moritz, and Philipp Sandner. "The Impact of Crypto Currencies on Developing Countries." FSBC Working Paper. December 2019. http://explore-ip.com/2019_The-Impact-of-Crypto-Currencies-on-Developing-Countries.pdf.

"Jedi." *Wikipedia*. Accessed August 31, 2022. https://en.wikipedia.org/wiki/Jedi.

Mueller, Benito. *Equity in Climate Change: The Great Divide*. Oxford: Oxford Institute for Energy Studies, 2002. https://www.oxfordenergy.org/wpcms/wp-content/uploads/2011/03/EV31-EquityinGlobalClimateChangeTheGreatDivide-BMuller-2002.pdf.

Nichols, Nichelle. *Beyond Uhura: Star Trek and Other Memories*. London: Boxtree, 1995.

"Nyota Uhura." *Wikipedia*. Accessed August 31, 2022. https://en.wikipedia.org/wiki/Nyota_Uhura.

Thanki, Nathan. "What Does Greta Thunberg's Call for Equity Mean?" *Climate Home News*. February 10, 2019. https://www.climatechangenews.com/2019/10/02/greta-thunbergs-call-equity-mean.

Whelan, Tensie. *US Corporate Boards Suffer from Inadequate Expertise in Financially Material ESG Matters*. NYU Stern Center for Sustainable

Business. January 2021. https://www.stern. nyu.edu/sites/default/files/assets/documents/ US%20Corporate%20Boards%20Suffer%20 From%20Inadequate%20%20Expertise%20 in%20Financially%20Material%20ESG%20 Matters.docx%20%282.13.21%29.pdf.

CONCLUSION

Cambridge Carbon Footprint. "Take Fewer Baths and Shorter Showers." https://cambridgecarbon-footprint.org/actions/fewer_baths_shorter_ showers.

Resources

All We Can Save (Ayana E. Johnson and Katherine K. Wilkinson)

Atomic Habits (James Clear)

Billions and Billions (Carl Sagan)

Blockchain Revolution (Don and Alex Tapscott)

Electrify (Saul Griffith)

How the World Really Works (Vaclav Smil)

How to Avoid a Climate Disaster (Bill Gates)

Invent & Wander (Jeff Bezos)

Mission Economy (Mariana Mazzucato)

Revolutionary Power (Shalanda Baker)

Sapiens (Yuval Noah Harari)

Speed and Scale (John Doerr)

Termination Shock (Neal Stephenson)

The Great Derangement (Amitav Ghosh)

The Ministry of the Future (Kim Stanley Robinson)

The Nutmeg's Curse (Amitav Ghosh)

The Overstory (Richard Powers)

The Power of Ritual (Casper ter Kuile)

The Prize (Daniel Yergin)

The Sum of Us (Heather McGee)

The Tree Line (Ben Rawlence)

The Wizard & the Prophet (Charles Mann)

Under a White Sky (Elizabeth Kolbert)

Why Trust Science (Naomi Oreskes)

Acknowledgments

I would like to thank the following people, without whom I would not have been able to complete my book writing journey.

First, I want to thank my editors, Rochelle Melander, Stephanie Kilen, and Steve Johnson, who over a three-year period helped me transform a loosely connected set of ideas into a stunningly coherent book.

Second, I want to thank Sarah Lahay for being the greatest cover and interior designer I could ever imagine, and collaborating to transform my words into a visual work of art.

Third, I want to thank everyone on the New Degree Press team for helping me navigate the publishing landscape. Special thanks to Eric Koester, for envisioning and creating this amazing platform for aspiring new writers; and Brian Bies, for helping to simplify the marketing and publishing process.

Finally, I especially want to thank my early supporters who purchased my book and believed in me as an author. None of this would have been possible without you:

EARLY SUPPORTERS

Silvia Aldana

Kirk Coburn

Lucinda Crabtree

Sarah Fitts

Russell Frisby

Elizabeth Hackenson

Pari Kasotia

David Owens

John Parsons

Emily Robichaux

Natalie Roy

Jigar Shah

Stuart Smits

Rachel Smucker

Karluss Thomas

Charles Walton

Lina Williams

ENDNOTE

Please note that any reference to groups or individuals is not an endorsement of their opinions or viewpoints.

To those who mimic the concept or framework of this book, my only request is that you honor the source of inspiration, as I have attempted to do with others.

Appendix

THE COASTAL CREED

This appendix is dedicated to National Clean Beaches Week (NCBW), a coastal creed I created, that was endorsed by the White House and over 150 governors, mayors, and counties. The creed became a House and Senate resolution that was passed by unanimous consent in the 110th United States Congress.

PROCLAMATIONS AND RESOLUTIONS (U.S. GOVERNORS, MAYORS, AND COUNTY COMMISSIONS)

Alabama (Governor Bob Riley)
    Mobile (Mayor Sam Jones)

Alaska (Governor Sarah Palin)

<u>California</u> (Governor Arnold Schwarzenegger)
 Long Beach (Mayor Bob Foster)
 Los Angeles (Mayor Antonio R. Villaraigosa)
 Milpitas (Mayor Jose Esteves)
 San Diego (Mayor Jerry Sanders)
 San Francisco (Mayor Gavin Newsom)
 San Rafael (Mayor Albert J. Boro)
 Santa Barbara (Mayor Marty Blum)
 Santa Monica (Mayor Richard Bloom)
 Marin County Board of Supervisors
 Mendocino County Board of Supervisors
 Santa Cruz County Board of Supervisors
 Sonoma County Board of Supervisors
 Ventura County Board of Supervisors

<u>Connecticut</u> (Governor M. Jodi Rell)
 Bridgeport (Mayor John Michael Fabrizi)
 Fairfield (Mayor Kenneth Flatto)

<u>Delaware</u> (Governor Ruth Ann Minner)
 Sussex County Council

<u>Florida</u> (Charlie Crist)
 Clearwater (Mayor Frank Hibbard)
 Delray Beach (Mayor Rita Ellis)
 Jacksonville (Mayor John Peyton)
 Miami (Mayor Manuel A. Diaz)
 Naples (Mayor Bill Barnett)

Orlando (Mayor Buddy Dyer)

Palm Beach (Mayor Jack McDonald)

Sarasota (Mayor Richard Clapp)

St. Petersburg (Mayor Rick M. Baker)

Tampa (Mayor Pam Iorio)

Venice (Mayor Fred Hammett)

Bay County Board of Commissioners

Brevard County Board of Commissioners

Broward County Board of Commissioners

Charlotte County Board of Commissioners

Collier County Board of Commissioners

Jacksonville City Council

Flagler County Board of Commissioners

Indian River County Board of Commissioners

Lee County Board of Commissioners

Manatee County Board of Commissioners

Martin County Board of Commissioners

Miami-Dade Office of the Mayor and Board of County Commissioners

Monroe County Board of Commissioners

Pasco County Board of Commissioners

Pinellas County Board of Commissioners

Volusia County Council

Walton County Board of Commissioners

Georgia

Camden County Board of Commissioners

Hawaii (Governor Linda Lingle)
    Maui County Council, City of Honolulu

Illinois (Governor Rod R. Blagojevich)
    Highland Park (Mayor Michael D. Belsky)
    Lake County Board

Indiana (Governor Mitch Daniels)
    Gary (Mayor Rudy Clay)
    Michigan City (Mayor Chuck Oberlie)
    Porter County Board of Commissioners

Louisiana (Governor Bobby Jindall)
    Kenner (Mayor Edmond J. Muniz)
    New Orleans (Mayor C. Ray Nagin)
    Jefferson Parish Council

Maine (Governor John Elias Baldacci)
    Waldo County Commissioners

Maryland (Governor Martin O'Malley)
    Baltimore (Mayor Sheila Dixon)

Massachusetts (Governor Deval Patrick)
    Boston (Mayor Thomas M. Menino)

Michigan (Governor Jennifer Granholm)
    Allegan County Board of Commissioners

Grand Traverse County Board of Commissioners
Macomb County Board of Commissioners
Wayne County Commission

Minnesota (Governor Tim Pawlenty)
Saint Paul (Mayor Chris Coleman)

Mississippi (Governor Haley Barbour)
Biloxi (Mayor A.J. Holloway)

New Hampshire (Governor John Lynch)

New Jersey (Governor Jon Corzine)
Atlantic County Board of Freeholders

New York (Governor Eliot Spitzer)
Buffalo (Mayor Byron Brown)
New York City (Mayor Michael R. Bloomberg)
Chautauqua County Legislature
Nassau County Executive
Niagara County Legislature

North Carolina (Mike Easley)
Kill Devil Hills (Mayor Ray Sturza)
Kitty Hawk (Mayor Clifton Perry)
Southern Shores (Mayor Don Smith)
Wilmington (Mayor Bill Saffo)
Carteret County Board of Commissioners

    Currituck County Board of Commissioners
    New Hanover County Board of Commissioners

<u>Ohio</u>
    Cleveland (Mayor Frank Jackson)
    Erie County Board of Commissioners
    Lake County Board of Commissioners
    Lorain County Board of Commissioners

<u>Oregon</u> (Governor Ted Kulongoski)
    Curry County Board of Commissioners

<u>Pennsylvania</u> (Governor Edward G. Rendell)
    Erie (Mayor Joseph Sinnott)

<u>Rhode Island</u> (Governor Donald L. Carcieri)
    Providence (Mayor David N. Cicilline)

<u>South Carolina</u> (Governor Mark Sanford)
    Beaufort (Mayor William Rauch)
    Charleston (Mayor Joseph P. Riley)
    Horry County Council

<u>Texas</u> (Governor Rick Perry)
    Corpus Christi (Mayor Henry Garrett)
    Galveston (Mayor Lyda Ann Thomas)
    Calhoun County Board of Commissioners
    Matagorda County Commissioners' Court
    San Patricio County Judge

<u>Virginia</u> (Governor Tim Kaine)
 Hampton (Mayor Ross A. Kearney)
 Norfolk (Mayor Paul D. Fraim)
 Portsmouth (Mayor James W. Holley)
 Virginia Beach (Mayor Meyera E. Oberndorf)
 Northampton County Board of Supervisors

<u>Washington</u> (Governor Christine Gregoire)
 Gold Beach (Mayor Karl Popoff)
 Seattle (Mayor Greg Nickels)
 Grays Harbor County Board of Commissioners
 Jefferson County Board of Commissioners
 Pacific County Board of Commissioners

<u>Washington D.C.</u> (Mayor Adrian Fenty)

Wisconsin (Governor Jim Doyle)
 Green Bay (Mayor James J. Schmitt)
 Milwaukee (Mayor Tom Barrett)
 Ozaukee County Board of Supervisors
 Sheboygan County Board Chairman

H. Res. 186

In the House of Representatives, U. S.,

June 27, 2007.

Whereas coastal areas produce 85 percent of all United
States tourism dollars and are the leading tourism destination in America;

Whereas the National Oceanic and Atmospheric Administration reports that over 50 percent of the population of the
United States lives in coastal counties;

Whereas according to the National Oceanic and Atmospheric
Administration, the beaches in these coastal counties provide recreational opportunities for numerous Americans
and their families who, together with international tourists, make almost 2 billion trips to the beach each year
to fish, sunbathe, boat, swim, surf, and bird-watch;

Whereas according to the Army Corps of Engineers, United
States beaches are a critical driver of the American economy and its competitiveness in the global economy;

Whereas beaches represent a critical part of our natural heritage and a beautiful part of the American landscape;

Whereas beaches are sensitive ecosystems, susceptible to degradation and alteration from pollution, sea level rise, natural forces, untreated sewage, and improper use;

Whereas members of the government, the private sector, and
nongovernmental organizations, along with citizen volun-

2

teers, have worked diligently to clean up and protect our beaches over the years;

Whereas according to the United States Geological Survey, great strides have been made in understanding the science of watersheds and the connections between inland areas and coastal waters, and science-based policy should be developed that is commensurate with this knowledge; and

Whereas a 7-day week commencing in June, and including July 5, will be observed each year as National Clean Beaches Week: Now, therefore, be it

Resolved, That the House of Representatives—

(1) supports the goals and ideals of National Clean Beaches Week;

(2) recognizes the value of beaches to the American way of life and the important contributions of beaches to the economy, recreation, and natural environment of the United States;

(3) encourages all Americans to work to keep beaches, which are a critical part of the natural heritage of the United States, safe and clean for the continued enjoyment of the public;

(4) expresses a renewed appreciation for the beaches of the United States and an invigorated effort to protect them with updated, integrated policy; and

(5) encourages individuals to engage in activities during National Clean Beaches Week to encourage stewardship and volunteerism along our coastlines.

Attest:

Clerk.

110TH CONGRESS
1ST SESSION

S. RES. 243

Supporting the goals and ideals of National Clean Beaches Week and the considerable value of beaches and their role in American culture.

IN THE SENATE OF THE UNITED STATES

JUNE 20, 2007

Mr. LAUTENBERG (for himself, Mr. MARTINEZ, Mr. LIEBERMAN, Mrs. DOLE, Ms. STABENOW, Mr. STEVENS, Mr. BIDEN, Mr. BURR, Mr. LEVIN, Ms. MURKOWSKI, Mr. KERRY, Ms. SNOWE, Ms. LANDRIEU, Mr. LOTT, Mr. MENENDEZ, Mr. DURBIN, Mr. WYDEN, Mr. FEINGOLD, Mr. CARDIN, Mr. CARPER, and Ms. CANTWELL) submitted the following resolution; which was considered and agreed to

---

RESOLUTION

Supporting the goals and ideals of National Clean Beaches Week and the considerable value of beaches and their role in American culture.

Whereas coastal areas produce 85 percent of all United States tourism dollars and are the leading tourism destination in America;

Whereas over 50 percent of the population of the United States lives in coastal counties;

Whereas the beaches in these coastal counties provide recreational opportunities for numerous Americans and their families who, together with international tourists, make almost 2,000,000,000 trips to the beach each year to fish, sunbathe, boat, swim, surf, and bird-watch;

Whereas beaches are a critical driver of the American economy and its competitiveness in the global economy;

Whereas beaches represent a critical part of our natural heritage and a beautiful part of the American landscape;

Whereas beaches are sensitive ecosystems, susceptible to degradation and alteration from natural forces, sea level rise, pollution, untreated sewage, and improper use;

Whereas members of the Government, the private sector, and nongovernmental organizations, along with citizen volunteers, have worked diligently to clean up and protect our beaches over the years;

Whereas great strides have been made in understanding the science of watersheds and the connections between inland areas and coastal waters;

Whereas the Federal Government should develop science-based policies that are commensurate with that knowledge; and

Whereas a 7-day week, commencing in June and including July 5, will be observed as National Clean Beaches Week: Now, therefore, be it

1   *Resolved,* That the Senate—

2   (1) supports the goals and ideals of National

3   Clean Beaches Week;

4   (2) recognizes the value of beaches to the Amer-

5   ican way of life and the important contributions of

6   beaches to the economy, recreation, and natural en-

7   vironment of the United States; and

1   (3) encourages Americans to work to keep
2   beaches safe and clean for the continued enjoyment
3   of the public and to engage in activities during Na-
4   tional Clean Beaches Week that foster stewardship,
5   healthy living, and volunteerism along our coastlines.

○

THE WHITE HOUSE

WASHINGTON

June 25, 2007

I send greetings to those observing National Clean Beaches Week.

Our Nation's beaches provide magnificent scenery, diverse plant and animal life, and many opportunities to enjoy healthy outdoor recreation. All Americans share a responsibility to ensure these areas remain vibrant resources and places of inspiration and wonder for generations to come. This week is a chance to celebrate our rich natural heritage and underscore the importance of conserving America's marine environment.

My Administration remains committed to protecting our air, water, and land by pursuing innovative environmental initiatives and by continuing to enforce environmental laws. By working together, we can help build a more beautiful America.

I appreciate the Clean Beaches Council and all those who are committed to protecting and improving our country's beaches. Your efforts set an excellent example of responsible citizenship and help build a brighter future for our children and our Nation.

Laura and I send our best wishes.

2005 STATE OF THE BEACH REPORT: BACTERIA AND SAND

A NATIONAL CALL TO ACTION

JULY 2005

ACKNOWLEDGEMENTS

ABOUT CBC
The mission of CBC is to preserve America's coastal heritage through public and community awareness and voluntary participation in programs that will ensure a legacy of clean beaches and healthy coastal environments for generations to come. The CBC envisions a global movement to protect earth's coastal regions while promoting sustainable use and habitation. The Council will be a recognized leader and contributor in that movement. Visit us at www.cleanbeaches.org

ACKNOWLEDGEMENTS
CBC wishes to acknowledge the in-kind support of the Law Firm Bryan Cave, LLP, the U.S. Geological Survey, original sponsors of the National Clean Beaches Week Resolution: Representative Clay Shaw (R-FL), Representative Frank Pallone (D-NJ), and the more than 180 million Americans who love the beach and make more that 2 billion visits every year. We would also like to thank our team that worked on this report.

TABLE OF CONTENTS

Message from the President

When the Clean Beaches Council was founded in 1998, our mission included increasing public awareness of and literacy about the fragile coastal ecosystem. This would help ensure a legacy of clean and healthy beaches for future generations. Today, both of these goals continue to be the driving force behind the Council s campaigns and have motivated our efforts to produce the *2005 State of the Beach Report: Bacteria and Sand.*

This report is meant to enhance the science literacy of the beach going public about the state of our beaches today. The report examines the scientific evidence for bacteria contamination in beach sands. *Bacteria and Sand* is designed to actively educate the public about potential environmental contamination at the beach while emphasizing that these conditions are often affected by human actions. In fact, sand contamination may often be the result of careless human litter. Consider the following scenario:

A beachgoer leaves food waste on the beach, which attracts birds/wildlife. The birds/wildlife defecates on the beach, resulting in sand contamination. As the tide interacts with the sand, bacteria are released into the water, causing water contamination. Human exposure to these contaminated waters creates a health risk.

In this scenario, sand contamination is preventable. *Bacteria and Sand* aims to curb this type of behavior through public science literacy, recommending the Leave No Trace principle for the beach. Beachgoers need to understand that the only footprint they leave should be an imprint in the sand.

However, solving the problems raised in *Bacteria and Sand* will require more than a public awareness and literacy campaign about litter. Scientists believe that there may be many unknown routes of sand contamination that have yet to be addressed. Hence, this is just the start of a journey to better understand the nature of bacteria and sand.

We hope you will join us.

Sincerely,

Walter L. McLeod
President

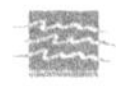

Executive Summary

Beaches serve an important role in the U.S. economy and the natural environment. Each year approximately 180 million Americans visit our nation s ocean, gulf and inland beaches. Coastal recreation and tourism are estimated to contribute over $640 billion annually to the U.S. economy (85% of all U.S. tourist revenues). At the same time, beaches provide a crucial habitat for marine life and coastal species. Moreover, beaches often adjoin wetland and coral reef ecosystems, both recognized for their ecological value.

Yet, our coastlines are threatened. The growing population and industry along both the East and West coasts as well as the Great Lakes are increasingly polluting our beaches. This pollution comes in the form of beach litter, sewage overflows, storm water runoffs, agriculture and industrial discharges, and recreational activities, such as boating. For instance, from 2002 to 2003, the number of beach closings rose 51% to a staggering 18,000 reported closings nationwide.[1] While some of this increase can be attributed to more accurate and frequent testing of recreational water, the growing impact of humans cannot be ignored.

Responding to this emerging threat, environmental agencies and health departments regularly test the swimming water for the presence of fecal indicator bacteria such as *E.coli* and *enterococci*. High densities of indicator bacteria in the water often prompt state or local agencies to close beaches until a safe level is reached.

Not until recently, however, have studies investigated the presence of indicator bacteria (*E. coli*) in the sand itself. In 1994, researchers first began to notice that concentrations of *E. coli* bacteria were much higher in the sand and in shallow water than in the deeper water. A 2003 United States Geological Survey study found that indicator bacteria levels in sand averaged 5 - 10 times higher than levels in adjacent swimming waters. Scientists suspect that while many pathogens find water a less hospitable environment, bacteria harbored in the sand may persist longer than in the water because they adhere to sediment particles, unlike free bacteria in water. At present, however, there is no conclusive link between indicator bacteria in sand and human health.

The Clean Beaches Council's *2005 State of the Beaches Report: Bacteria and Sand* is an in-depth look at the current research addressing the issue of bacteria and sand. It is intended to catalyze action on this emerging field of science and improve the health and quality of America s beaches.

This report also serves as a call to action for our leaders in government, the scientific community and the beach-going public to take the following actions:

Call to Action 1: Establish a national science program to address the geological/biological processes between bacteria and sand, including epidemiological studies on the potential link to human health.

Call to Action 2: Establish a global network of researchers devoted to advancing the science of geological/biological processes of bacteria and sand, and their subsequent impacts on human health.

Call to Action 3: Establish a national coastal corps network of beach goers, committed to increasing beach science literacy through education, outreach and action.

To help us advance the call to action, please visit us online at www.cleanbeaches.org or contact us by email at info@cleanbeaches.org.

Executive Summary

Beaches serve an important role in the U.S. economy and the natural environment. Each year approximately 180 million Americans visit our nation s ocean, gulf and inland beaches. Coastal recreation and tourism are estimated to contribute over $640 billion annually to the U.S. economy (85% of all U.S. tourist revenues). At the same time, beaches provide a crucial habitat for marine life and coastal species. Moreover, beaches often adjoin wetland and coral reef ecosystems, both recognized for their ecological value.

Yet, our coastlines are threatened. The growing population and industry along both the East and West coasts as well as the Great Lakes are increasingly polluting our beaches. This pollution comes in the form of beach litter, sewage overflows, storm water runoffs, agriculture and industrial discharges, and recreational activities, such as boating. For instance, from 2002 to 2003, the number of beach closings rose 51% to a staggering 18,000 reported closings nationwide.[1] While some of this increase can be attributed to more accurate and frequent testing of recreational water, the growing impact of humans cannot be ignored.

Responding to this emerging threat, environmental agencies and health departments regularly test the swimming water for the presence of fecal indicator bacteria such as *E.coli* and *enterococci*. High densities of indicator bacteria in the water often prompt state or local agencies to close beaches until a safe level is reached.

Not until recently, however, have studies investigated the presence of indicator bacteria (*E. coli*) in the sand itself. In 1994, researchers first began to notice that concentrations of *E. coli* bacteria were much higher in the sand and in shallow water than in the deeper water. A 2003 United States Geological Survey study found that indicator bacteria levels in sand averaged 5 - 10 times higher than levels in adjacent swimming waters. Scientists suspect that while many pathogens find water a less hospitable environment, bacteria harbored in the sand may persist longer than in the water because they adhere to sediment particles, unlike free bacteria in water. At present, however, there is no conclusive link between indicator bacteria in sand and human health.

The Clean Beaches Council' s *2005 State of the Beaches Report: Bacteria and Sand* is an in-depth look at the current research addressing the issue of bacteria and sand. It is intended to catalyze action on this emerging field of science and improve the health and quality of America s beaches.

This report also serves as a call to action for our leaders in government, the scientific community and the beach-going public to take the following actions:

Call to Action 1: Establish a national science program to address the geological/biological processes between bacteria and sand, including epidemiological studies on the potential link to human health.

Call to Action 2: Establish a global network of researchers devoted to advancing the science of geological/biological processes of bacteria and sand, and their subsequent impacts on human health.

Call to Action 3: Establish a national coastal corps network of beach goers, committed to increasing beach science literacy through education, outreach and action.

To help us advance the call to action, please visit us online at www.cleanbeaches.org or contact us by email at info@cleanbeaches.org.

Chapter 1: Introduction

In the United States, where tourism is the largest industry, employer, and foreign revenue earner, coastal states earn 85% of all U.S. tourism revenues.[3] Besides being vital to the American economy, beaches are also varied and rich ecosystems. They are home to a number of plant and animal species, and also serve as fertile breeding grounds. Beneath the tranquil veneer of sand and surf is an ever-changing community of flora and fauna striving to maintain their habitats.

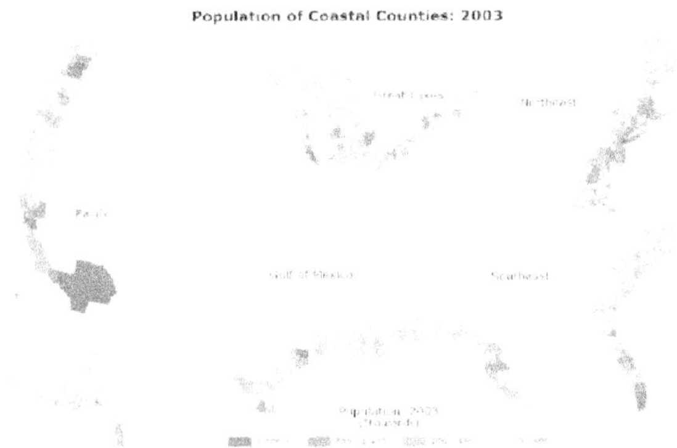

Fig. 1. 2003 Coastal County Population (Source: U.S. Census Bureau)[4]

Coastal counties have become increasingly populated and are among the most crowded and developed in the nation today. In 2003, over 153 million Americans lived in the 673 coastal counties along the Atlantic and Pacific Oceans, Gulf of Mexico, and Great Lakes (see Figure 1), representing over 53 % of the total U.S. population.[5]

An unfortunate consequence of this development has been the slow deterioration of these beaches, as population densities in coastal counties increase with each year. Coastal counties (excluding Alaska) average 300 persons per square mile, much higher than the national average of 98 persons per square mile.[6] A study found that the number of closings and advisories at U.S. oceans and Great Lakes beaches in 2003 rose 51% from 2002.[7] While beachgoers are just beginning to understand traditional beach issues, such as erosion and water pollution, sand contamination is an important emerging issue. Scientists are voicing concern and raising questions regarding sand-borne indicator bacteria as a source of beach pollution.

About Beaches

How are beaches formed?

Over millions of years, sand and gravel from weathered rocks and other eroded surfaces are washed to sea by streams and rivers. Most sediment is suspended in seawater and is carried to the shoreline by two separate processes known in combination as littoral drift.[8] In the first process, the long shore current flowing parallel to the shore carries this sedimentary material along the coast. This parallel flow is created by the action of waves breaking at an angle on the shore. Long shore transport can deliver up to a million cubic yards of sediment annually to a single beach.[9] The second process involves the deposition of sand on shore by oscillating waves moving in a direction perpendicular to the shore. As the waves alternatively crash into and recede from the shore, sand is progressively pushed along the beach edge, resulting in the formation of a beach.

What are healthy and sick beaches?

In addition to the presence or absence of pollution, there are other factors that may affect the health of a beach. For instance, researchers in New Zealand concluded that the rate at which a beach is able to dry out may be indicative of its potential health impacts on humans and depends on the slope of the beach.[10] Figure 2 compares a steep beach with a flat beach. With the water table being low in a steep beach, hydrostatic pressure is high and this enables the water to flow more quickly through the sand. These researchers suggest that this helps the beach to dry out faster and makes it less prone to bacteria contamination.

On a low-sloped beach however, the water takes more time to reach the sea, thus extending the drying out period. Under certain conditions such a beach will be more prone to erosion, especially during a major storm event.

Fig. 2. Healthy and sick beaches[11]

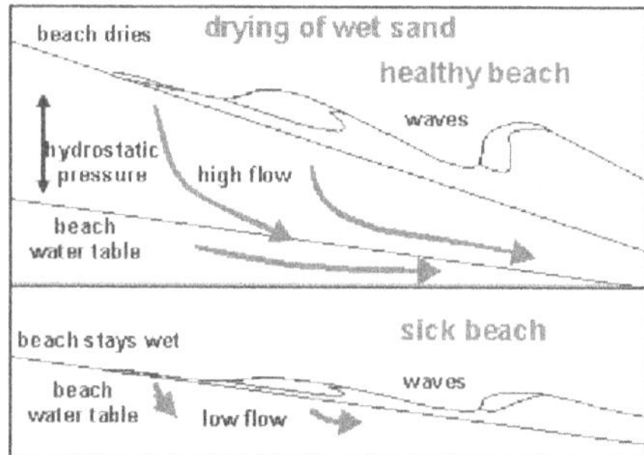

One sign of a sick beach is the presence of crusting bacteria. On a sick beach polluted by fine particles or bacteria, there are not enough air channels and pockets available to let air replace the lost water. As the moisture and wetness evaporate from the drying sand, the subsequent vacuum that builds up pulls up the water table. The absence of air causes the water with solutes to rush in and fill up the vacuum. This action results in the formation of a crust that may contain high levels of bacteria and salt, causing the sand grains to cake together. According to these researchers, the presence of crusting bacteria may indicate a sickly beach and a potential health hazard.

Traditional Environmental Problems at Beaches

Beach Litter

While most Americans take care to enjoy our beaches respectfully, marine debris continues to be a problem for

our coasts. In 2003, 450,000 volunteers picked up 7.6 million pounds of trash from 16,250 miles on shore and under water.[12] Despite repeated warnings on the destruction beach litter can cause to the surrounding flora and fauna, many continue to think of beaches as convenient trash cans and ashtrays.

Water Quality

Water quality problems have long plagued our nation s coasts. With increasing crowds, most beaches now suffer from the presence of fecal coliforms, a bacterial indicator for human pathogens that may be present in the water. As overextended health departments strive to put adequate water monitoring programs in place, there is still a great deal of debate among the scientific community regarding the best possible way to keep beachgoers safe and healthy. The main sources of contamination for recreational waters are sewage overflows, agricultural and industrial discharges, storm water runoffs, the presence of birds and animals, and the beach users themselves.

How can a beachgoer benefit from this report?

This report is meant to enhance the science literacy of the beach going public about the state of our beaches today. The report examines the science of bacteria contamination in beach sands. As more studies reveal, our focus on water quality may have caused us to overlook the possibility of contaminated sand. Bacteria do exist in sand, and in some cases, survive better in sand than in water. Preliminary studies suggest that sand may actually serve as a reservoir for fecal bacteria and could possibly be a source of contamination to water. However, as this report goes on to discuss, only further research will reveal the true nature of the relationship between sand and bacteria.

Chapter 2: All About Sand

A Brief Introduction to Sand

What is sand?

Sand is normally a reflection of nearby bedrock. It is a result of millions of years of mechanical and chemical breakdown of rocks. The composition of sand is largely dependent on the source material. For instance, the sand on some Hawaiian beaches is often composed of volcanic rock fragments, volcanic glass, and cinder from volcanic eruptions. In contrast, sediment found on many of the beaches of southern California comes from granites, a common rock which consists of quartz plus tiny amounts of mica, feldspar, and hornblende. The white sand beaches of Mexico are mainly composed of shell fragments and other organic material. Individual sand grains are the size of table salt grains (less than one millimeter in diameter) and resemble miniature gemstones when magnified.

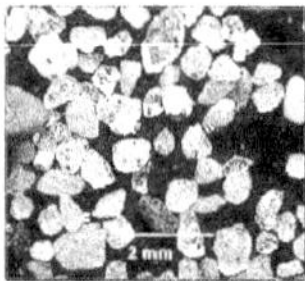
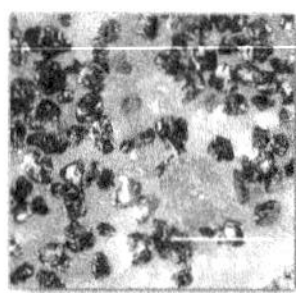

Lake Powell (Utah)   California Black Sand

Cabo, Baja Mexico Sand   Waianapanapa, Maui Sand

Fig. 3. Sands from different beaches[13] (' Microbus)

What are most common beach sand issues?

Beaches play a very important role as habitat to a number of plants and animals. They serve as breeding grounds, shelters, sources of food for marine life, and are home to a number of endangered species.[14] The overcrowding of beaches has led to large-scale destruction of some of these habitats and has reduced their ability to adapt to environmental changes. Development, climate change, and commerce have all played a part in increasing the pressure on beach ecosystems.

Beach erosion, defined simply as the loss of sand, is the direct consequence of a combination of different factors. These include building coastal developments too close to the beach and a gradual rise in sea level.[15] In a move to compensate for the lost sand and prevent further losses, communities often undertake shoreline armoring and beach renourishment projects. A sand issue that has caused considerable debate is the slow but sure disappearance of sand dunes.

How are sand dunes formed? Why are they important?

Sand dunes are defined as natural or artificial ridges or mounds of sand landward of the beach.[16] Dunes are geological landforms that are formed when there is a combination of abundant sand, strong winds, and sparse vegetation. Depending on proportions in which these three variables are present, different dune types may result, including U-shaped dunes (*parabolic* dunes), linear or *longitudinal* dunes and *transverse* dunes. Places with minimal sand supply, weak winds and/or abundant vegetation are not conducive to the formation of sand dunes.

Coastal dunes are of great ecological importance. They support the habitats of a number of rare and unique plant and animal species. They also play a vital role in providing a natural barrier against flooding and erosion. In the United States, the dunes on the shores of Lake Michigan are well known tourist attractions, bringing in more than half a million people each year. These dunes, created some 10,000 years ago, comprise the largest concentration of freshwater sand dunes in the world and support more unique species and communities than any other part of the Great Lakes system. Recently, the existence of these dunes has been severely threatened from continuous strip mining and the dunes continue to disappear at a rapid rate even though various anti-mining laws have been passed.[17]

Bacterial Contamination: Origins and Monitoring

What is the historical view of bacteria and beaches?

Pollution at beaches is typically regarded as a direct consequence of industrial, agricultural, or municipal activities that result in the bacterial contamination of water. Recreational waters are tested for the presence of fecal indicator bacteria. Fecal indicator bacteria, which include *E. coli* and *enterococci*, are abundant in feces and so their presence indicates that the water has been contaminated. With the number of beach advisories and closings seeming to increase with every passing summer, beachgoers have become fairly aware of the health hazards associated with swimming in contaminated waters. Human wastes are generally considered the main source of fecal contamination. Other sources, such as beach sand, have been overlooked as a source of fecal bacteria.[18] However, recent studies are causing a change in perception. Results indicate:
- There are indirect sources of indicator bacteria, such as sand and plants, that may significantly affect recreational water quality.
- Indicator bacteria seems to be commonly found in beach sand, at levels sometimes greater than water.

How is beach water monitored?

The main objective of beach monitoring is to evaluate the presence and levels of fecal indicator bacteria in recreational water. Levels that exceed defined thresholds indicate that the beach water is contaminated and may be unfit for swimming. The levels of bacteria in the water are measured in colony forming units (CFUs) per 100 milliliters (mL) of water. A level of 100 CFU/100 mL would mean there are about 100 cells of these bacteria in every half-cup of water.

What are indicator bacteria? Are they harmful?

It is difficult, time-consuming, and expensive to test for all the potential pathogens in water. Therefore, the United States Environmental Protection Agency (EPA) recommends testing for fecal indicator bacteria. Fecal indicator bacteria are easy to analyze, relatively safe to handle, and are usually present when enteric (fecal) pathogens are present. Fecal indicator bacteria are microbes that also live and multiply in the digestive tract of humans and warm-blooded animals, but they normally do not cause disease.[19] If these indicator bacteria are present, some amount of fecal material in the water is likely. A higher level of indicator bacteria signifies a greater level of contamination from fecal matter and a greater chance that pathogenic microbes being present, thus presenting a health risk to swimmers from waterborne illness.

The *E. coli* used to measure water quality does not cause illness in humans. There are only a few types of *E. coli* pathogens that cause outbreaks of disease. The pathogenic *E. coli* occur quite rarely in feces. The regular (non-pathogenic) *E. coli* on the other hand is present in the gastrointestinal tract of humans and does not cause diseases in humans.[20]

An Emerging Emphasis on Sand

In 1994, Ghinsberg et. *al.* examined the occurrence of indicator bacteria in shallow surface sands of a Mediterranean beach. During the same time, studies in Lake Michigan revealed that indicator bacteria (*E. coli*) in foreshore sands were 5 to 10-fold higher than the adjacent lake water. It became known that while the concentrations of indicator bacteria remained roughly constant within the first 5 meters landward of the lake (*i.e.* in the sand), the counts rapidly decreased while moving into deeper water of the lake. Furthermore, indicator bacteria concentrations were found in places where there was absolutely no indication of any human contamination. So where was the bacteria coming from? All clues pointed in the direction of sand and scientists began to examine the relationship between sand and fecal bacteria more closely.

Analysis shows that bacteria are often unable to survive longer in water, finding sand more favorable and possibly increasing their numbers in sand environments. Bacteria harbored in the sand may persist longer than in the water because they adhere to sediment particles and are protected from the harmful effects of radiation from sunlight, unlike free bacteria in the water. [21]

Also, the movement of sand by wave action due to storms and commercial or recreational boating may increase the bacterial counts in the shallow waters near the shores, which could prove to be a health hazard for children playing there. [22]

When faced with a swimming health advisory or closing, many beachgoers will opt to remain at the beach and enjoy sand play activities. Hence, there is a legitimate interest among the scientific community to better understand how contamination in sand affects public health. Children are a particular concern because they spend a great deal of time playing and digging in the wet sand and are more susceptible to intestinal illness; the elderly and infirm would also be more susceptible to illnesses associated with contamination. [23]

In 2003, an intensive study by the United States Geological Survey at a freshwater beach in Chicago revealed some interesting details on the relationship between sand and indicator bacteria (*E. coli*). The study found that indicator bacteria levels in sand averaged 5-10 times higher than levels in adjacent swimming waters. During the course of this study, the city replaced the contaminated sand with truckloads of fresh sand; however within two weeks indictor bacteria levels were similar to those collected before sand removal. What surprised the scientists most was that these indicator bacteria remained consistently high regardless of beach water quality. Scientists determined that these bacteria were self-sustaining in sand and did not need any additional sources to help boost their counts.

The finding that indicator bacteria concentrations were higher in shallow water than in deeper water served to indicate that one source of contamination in beach waters may also be the surrounding sand and sediments. After completing their research, scientists were able to conclude that sand was a major contributor of fecal indicator bacteria in lake water. The following two studies profiled in this report demonstrate the research being done in this area. These studies clearly show the presence of indicator bacteria in freshwater sands and marine sands.

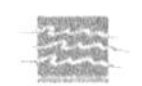

Chapter 3: Sand Contamination Studies

Study 1: Bacteria in Freshwater Sands

Study by: Elizabeth Wheeler Alm, Janice Burke and Anne Spain[24]

Title: Fecal indicator bacteria are abundant in wet sand at freshwater beaches[25]

Period: 2001 - 2002

Location: Six public bathing beaches along Lake Huron in St. Clair County, Michigan

Purpose of Study: The purpose of this study was to evaluate the potential of wet, freshwater sand to be a reservoir of intestinal bacteria. This was one of the first studies to address the survival and persistence of fecal bacteria in freshwater sands.

What is Known:

- Fecal contamination at bathing beaches can be hazardous to humans, because feces may contain bacteria, viruses, and protozoa that can be ingested and cause intestinal disease.
- With no additional input, levels of fecal indicators decline, in water. Several factors have been proposed to explain this decline including inactivation due to UV radiation,[26] and sunlight inactivation.[27] Additionally, fecal bacteria are also less able to obtain nutrients in water unlike those associated with the sediment particles.[28]
- Enteric bacteria have been known to exhibit longer survival times in sediment than in overlaying water.[29]

What the Study Confirmed:

- At each of the six beaches, fecal indicator bacteria were more abundant in sand than in water.
- Compared to water, enterococci counts in sand were 4—38 times higher and *E. coli* counts were 3—17 times higher.
- Enterococci counts were consistently low in water, but they were dramatically higher in the sand.
- The results of this study were consistent with work on freshwater beaches in England, where fecal indicator counts were an order of magnitude greater in sand than in the overlying water.
- The presence of fecal indicator bacteria in beach sand suggests that pathogenic bacteria of intestinal origin may also be present in the sand.

Results of Study

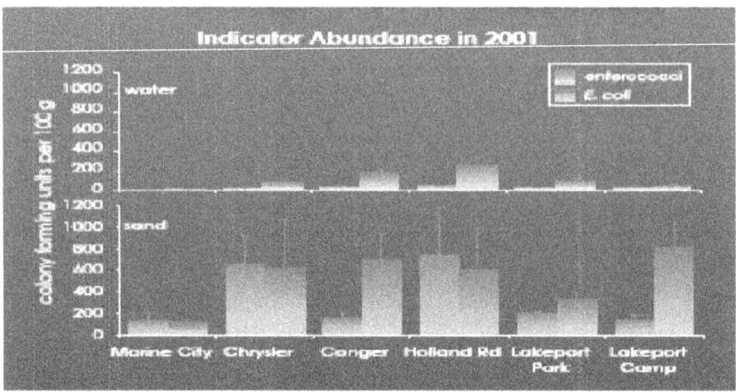

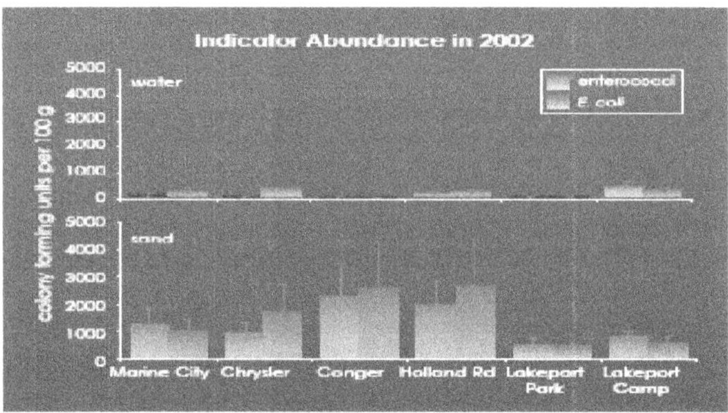

Printed in: Alm, E.W., J.M. Burke, E.L. Francis, and A.J. Matthews. 2002. Fecal Indicator Abundance Higher in Freshwater Bathing Beach Sand Than in Water Column. Great Lakes Beach Conference 2002. [on-line http://www.great-lakes.net/glba/pdf/Alm.pdf. Accessed 6/30/2005]

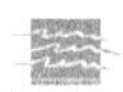

Study 2: Bacteria in Marine Sands

Study by: Tomoyuki Shibata*, Helena M. Solo-Gabriele*[30], Lora E. Fleming[31], Samir Elmir[32]

Title: Monitoring Marine Recreational Water Quality Using Multiple Microbial Indicators in an Urban Tropical Environment[33]

Period: March 2001 — August 2001

Location: Two Beaches in Miami-Dade County, Florida

Purpose of Study: This study does not look at the presence of indicator bacteria in sand specifically. Rather, the primary objective of the authors was to evaluate and compare multiple indicator microbes for two beach sites located in Miami, Florida. In the course of their study, they also describe the presence of these indicator bacteria in marine sands.

What is Known:

- The beach sand survey, conducted at Beach A, focused on collecting samples from the swash zone of the beach. The swash zone is the depth zone in which sediments are disturbed by wave action near the shoreline[34].
- Studies have shown that fecal coliform, *E. coli*, and enterococci are found in the environment in the absence of a known sewage source of contamination and have been shown to multiply within warm tropical environments.

What the Study Confirmed:

- Results showed that concentrations of indicator microbes were highest during high tide with hot spots at various points along the shoreline. None of the offshore sampling points had detectable levels of indicator microbes. The waters around the Miami Seaquarium generally had low levels of microbes. These results suggest that the shoreline is the primary source of indicator microbes and this source is most pronounced during high tide when the water level reaches its highest point along the shore.
- The results suggest that there may be an additional source(s) of indicator microbes other than sewage.
- Data collected indicated that the source of indicator microbes comes from the shoreline, as evidenced by the higher indicator microbe concentrations as the shore was approached. The authors agree that one possible source of the microbes may be the sand itself.
- In the results from the sand survey, indicator microbes were observed in all sand samples collected from the shoreline at Beach A.
- The authors recommend that the role of sand be further evaluated on the survival and the multiplication of the indicator microbes at this particular site.

Chapter 4: Conclusions

This report has only begun to scratch the surface of the complex relationship between sand and bacteria. There are a number of questions that are yet to be answered. Beach water quality monitoring in the United States has dramatically improved in the last decade with most states monitoring their beaches at least every two weeks. Unfortunately, the monitoring has been limited to testing only recreational waters.

The findings in this report challenge the traditional interpretation of bacteria contamination, occurrence and survival at recreational beaches. With more evidence pointing to sand as a likely source of bacteria contamination, perhaps it is time for a sea change in our national beach science policy. To keep beaches healthy for the millions who visit each year, it is imperative that we understand the biological and geological processes affecting bacterial contamination of sand and the potential risks to human health. At present, however, there is no conclusive link between bacteria in sand and human health.[35] In fact, we know very little about the complex nature of bacteria in the coastal environment.

Hence, CBC is calling on leaders in government, the scientific community and the beach going public to take the following actions:

Call to Action 1: Establish a national science program to address the geological/biological processes between bacteria and sand, including epidemiological studies on the potential link to human health.

Call to Action 2: Establish a global network of researchers devoted to advancing the science of geological/biological processes of bacteria and sand, and their subsequent impacts on human health.

Call to Action 3: Establish a national coastal corps network of beach goers, committed to increasing beach science literacy through education, outreach and action.

To help us advance the call to action, please visit us online at www.cleanbeaches.org or contact us by email at info@cleanbeaches.org.

Thank you for your support.

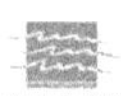

Appendix I: References

[1] Mark Dorfman, Testing the Waters 2003: A Guide to Water Quality at Vacation Beaches, a report for the Natural Resource Defense Council. < http://www.nrdc.org/water/oceans/ttw/titinx.asp>, (Accessed February, 2005).

[2] Whitman RL, Nevers MB, Foreshore sand as a source of Escherichia coli in nearshore water of a Lake Michigan beach. Appl Environ Microbiol. 2003 Sep;69(9):5555-62.

[3] National Oceanic and Atmospheric Administration, Ocean Facts on Coastal Tourism <http://www.yoto98.noaa.gov/facts/tourism.htm> (Accessed February, 2005).

[4] Kristen Crosset et al., Population Trends Along the Coastal United States 1998 — 2008, a report for the National Oceanic and Atmospheric Administration. <http://www.oceanservice.noaa.gov/programs/mb/pdfs/coastal_pop_trends_complete.pdf>, September 2004.

[5] Ibid.

[6] Ibid.

[7] Mark Dorfman, Testing the Waters 2003: A Guide to Water Quality at Vacation Beaches, a report for the Natural Resource Defense Council, < http://www.nrdc.org/water/oceans/ttw/titinx.asp>, (Accessed February, 2005).

[8] Ohio Office of Coastal Management, Beaches <http://www.dnr.state.oh.us/coastal/science/beaches.htm> (Accessed February 2005).

[9] California Coastal Commission California Coastal Resource Guide, <http://ceres.ca.gov/ceres/calweb/coastal/beaches.html,> (Accessed February, 2005).

[10] Floor Anthoni, SeaFriends <http://www.seafriends.org.nz>, New Zealand (Accessed March, 2005).

[11] Ibid.

[12] Oregon State Parks and Recreational Dept. International Beach Cleanup, <http://www.solv.org/programs/fall_beach_cleanup.asp> (Accessed February, 2005).

[13] Microbus, Take a Microscope Trip <http://www.microscope-microscope.org/microscope-home.htm>

[14] U.S. Commission on Ocean Policy.

[15] United States Geological Survey, Coastal Change <http://pubs.usgs.gov/circ/c1075/change.html> (Accessed March, 2005).

[16] NOAA Coastal Services Center <http://www.csc.noaa.gov/> (Accessed February, 2005).

[17] Lake Michigan Federation, An Advocate's Field Guide to Protecting Lake Michigan <http://www.greatlakes.org/field_guide/habitat_sand.asp>, 2004.

[18] Alm et. al. 2003. Fecal Indicator Bacteria Are Abundant in Wet Sand at Freshwater Beaches; Water Res 37:3978-3982.

[19] Dr. Dufour, *E.coli* and Public Health (2003) ; U.S. EPA; Lakeline 23:2: 13-15.

[20] Whitman RL, Nevers MB, Foreshore sand as a source of Escherichia coli in nearshore water of a Lake Michigan beach. Appl Environ Microbiol. 2003 Sep;69(9):5555-62.

[21] Ghosh et. al. Serovars of multi-antibiotic resistant Escherichia coli from the freshwater environs of Calcutta India. Microbiol Immunol 1991;35(4):273—88.

[22] Ibid.

[23] Ibid.

[24] Department of Biology, Central Michigan University

[25] Alm et. al. 2003. Fecal Indicator Bacteria Are Abundant in Wet Sand at Freshwater Beaches; Water Res 37:3978-3982.

[26] Y. Y. Chan & E. G. Killick; The effect of salinity, light and temperature in a disposal environment on the recovery of E.coli following exposure to ultraviolet radiation 1995; 29(5) 1373-77

[27] Davies-Colley RJ et. Al. Inactivation of faecal indicator microorganisms in waste stabilisation ponds: interactions of environmental factors with sunlight, Water Res 1999;33(5):1220— 30.

[28] Davies CM, Long JAH, Donald M, Ashbolt NJ. Survival of fecal microorganisms in marine and freshwater sediments; Appl Environ Microbiol 1995;61(5):1888—96.

[29]Gerba, C. P., & J. S. McLeod. 1976. Effect of sediments on the survival of Escherichia coli in marine waters. Appl. Environ. Microbiol. 32:114—120.

[30] Department of Civil, Architectural, & Environmental Engineering, University of Miami

[31]Rosensteil School of Marine and Atmospheric Science, University of Miami

[32] Miami-Dade County Health Department

[33] Shibata T., Solo-Gabriele H., Fleming L.E., Elmir S., 2004, Monitoring Marine Recreational Water Quality Using Multiple Microbial Indicators in an Urban Tropical Environment, 2004, *Water Research*, 38;3119-3131

[34]Marine Biology Web, Glossary of Marine Biology   <http://life.bio.sunysb.edu/marinebio/glossary.tuvwxyz.html> (Accessed February, 2005)

[35] "Microbial aspects of beach sand quality," Guidelines for Safe Recreational Water Enviroments. Volume I, (World Health Organization: Geneva, 2003), pp. 118-124

Appendix II: Other Acknowledgements

The Clean Beaches Council would also like to thank the following Mayors and Governors for proclaiming National Clean Beaches Week (June 27 — July 3, 2005) as well as Congressional Representatives for co-sponsoring a House Resolution Honoring the Week.

Mayors:

Mayor James K. Hahn (CA)
Mayor Richard M. Murphy (CA)
Mayor Gavin Newsom (CA)
Mayor Anthony A. Williams (DC)
Mayor Jeff Perlman (FL)
Mayor Jim Naugle (FL)
Mayor John Peyton (FL)
Mayor Bill Barnett (FL)
Mayor Buddy Dyer (FL)
Mayor John Mazziotti (FL)
Mayor Mary Anne Servian (FL)
Mayor Richard M. Baker (FL)
Mayor Pam Iorio (FL)
Mayor Dean Calamaras (FL)
Mayor Fland Sharp (FL)
Mayor Frank Hibbard (FL)
Mayor Mufi Hannemann (HI)
Mayor Lane Harrison (IL)
Mayor Joseph M. Stahura (IN)
Mayor C. Ray Nagin (LA)
Mayor Thomas M. Menino (MA)
Mayor Martin O Malley (MD)
Mayor Kwame M. Kilpatrick (MI)
Mayor Randy C. Kelly (MN)
Mayor A. J. Holloway (MS)
Mayor Spence H. Broadhurst (NC)
Mayor Michael R. Bloomberg (NY)
Mayor Jane Campbell (OH)
Mayor Richard E. Filippi (PA)
Mayor Joseph P. Riley (SC)
Mayor Mark S. McBride (SC)
Mayor Thomas Peeples (SC)
Mayor Lyda Ann Thomas (TX)
Mayor Henry Garrett (TX)
Mayor Todd W. Pearson (TX)
Mayor Ross A. Kearney (VA)
Mayor Greg Nickels (WA)
Mayor James J. Schmitt (WI)

Governors:

Governor Bob Riley (AL)
Governor Arnold Schwarzenegger (CA)
Governor M. Jodi Rell (CT)
Governor Ruth Ann Minner (DE)
Governor Jeb Bush (FL)
Governor Linda Lingle (HI)
Governor Kathleen Babineaux Blanco (LA)
Governor Jennifer M. Granholm (MI)
Governor Robert L. Ehrlich, Jr. (MD)
Governor Haley Barbour (MS)
Governor Richard J. Codey (NJ)
Governor John Lynch (NH)
Governor George E. Pataki (NY)
Governor Michael F. Easley (NC)
Governor Ted Kulongoski (OR)
Governor Donald L. Carcieri (RI)
Governor Mark Sanford (SC)
Governor Mark R. Warner (VA)
Governor Christine Gregoire (WA)
Governor Jim Doyle (WI)

Congressional Representatives:

Representative Neil Abercrombie (D-HI),
Representative Madeleine Z. Bordallo (D-GU)
Representative Henry E. Brown Jr. (R-SC)
Representative Ken Calvert (R-CA)
Representative Ed Case (D-HI)
Representative Michael N. Castle (R-DE)
Representative Susan A. Davis (D-CA)
Representative Sam Farr (D-CA)
Representative Mark Foley (R-FL)
Representative Vito Fossella (R-NY)
Representative Katherine Harris (R-FL)
Representative Steve Israel (D-NY)
Representative Connie Mack (R-FL)
Representative Mike McIntyre (D-NC)
Representative Jeff Miller (R-FL)
Representative Grace F. Napolitano (D-CA)
Representative Solomon P. Ortiz (D-TX)
Representative Frank Pallone (D-NJ)
Representative Jim Saxton (R-NJ)
Representative Clay E. Shaw Jr. (R-FL)
Representative Christopher H. Smith (R-NJ)

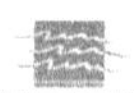

Appendix III: Works Cited

Alm et. al. 2003. Fecal Indicator Bacteria Are Abundant in Wet Sand at Freshwater Beaches; Water Res 37:3978-3982

Anthoni, Floor. SeaFriends <http://www.seafriends.org.nz> (Accessed March 2005)

California Coastal Commission California Coastal Resource Guide, http://ceres.ca.gov/ceres/calweb/coastal/beaches.html (Accessed February, 2005)

Chan Y. Y. et Killick E. G. The effect of salinity, light and temperature in a disposal environment on the recovery of E.coli following exposure to ultraviolet radiation 1995; 29(5) 1373-77

Crosset, Kristen. et al., Population Trends Along the Coastal United States 1998 — 2008, a report for the National Oceanic and Atmospheric Administration. http://www.oceanservice.noaa.gov/programs/mb/pdfs/coastal_pop_trends_complete.pdf, September 2004.

Davies-Colley RJ et. Al. Inactivation of fecal indicator microorganisms in waste stabilisation ponds: interactions of environmental factors with sunlight, Water Res 1999;33(5):1220— 30.

Davies CM, Long JAH, Donald M, Ashbolt NJ. Survival of fecal microorganisms in marine and freshwater sediments; Appl Environ Microbiol 1995;61(5):1888—96.

Dorfman, Mark Testing the Waters 2003: A Guide to Water Quality at Vacation Beaches, a report for the Natural Resource Defense Council, http://www.nrdc.org/water/oceans/ttw/titinx.asp>, (Accessed February, 2005).

Dufour, Dr. *E.coli* and Public Health (2003) ; U.S. EPA; Lakeline 23:2: 13-15

Gerba, C. P., et McLeod J.S.. 1976. Effect of sediments on the survival of Escherichia coli in marine waters. Appl. Environ. Microbiol. 32:114—120. Department of Civil, Architectural, & Environmental Engineering, University of Miami

Ghosh et. al. Serovars of multi-antibiotic resistant Escherichia coli from the freshwater environs of Calcutta India. Microbiol Immunol 1991;35(4):273 88

Lake Michigan Federation, An Advocate s Field Guide to Protecting Lake Michigan http://www.greatlakes.org/field_guide/habitat_sand.asp, 2004

Marine Biology Web, Glossary of Marine Biology http://life.bio.sunysb.edu/marinebio/glossary.tuvwxyz.html (Accessed February, 2005)

Microbus, Take a Microscope Trip <http://www.microscope-microscope.org/microscope-home.htm> (Accessed February, 2005)

National Oceanic and Atmospheric Administration, Ocean Facts on Coastal Tourism

http://www.yoto98.noaa.gov/facts/tourism.htm (Accessed February, 2005).

 NOAA Coastal Services Center <http://www.csc.noaa.gov/ (Accessed February, 2005)

Ohio Office of Coastal Management, Beaches http://www.dnr.state.oh.us/coastal/science/beaches.htm
     (Accessed February, 2005)

Oregon State Parks and Recreational Dept. Great Oregon Beach Cleanup,
     http://www.solv.org/programs/fall_beach_cleanup.asp (Accessed February, 2005)

Shibata T., Solo-Gabriele H., Fleming L.E., Elmir S., 2004, Monitoring Marine Recreational Water Quality
Using Multiple Microbial Indicators in an Urban Tropical Environment, 2004, Water Research, 38;3119-3131

United States Geological Survey, Coastal Change http://pubs.usgs.gov/circ/c1075/change.html> (Accessed
     March, 2005)

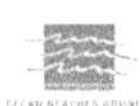